Physical Science

DIRECTED READING
WORKSHEETS ANSWER KEY

This book was printed with soy-based ink on acid-free recycled content paper, containing 10% POSTCONSUMER WASTE.

HOLT, RINEHART AND WINSTON

A Harcourt Classroom Education Company

Austin • New York • Orlando • Atlanta • San Francisco • Boston • Dallas • Toronto • London

Copyright © by Holt, Rinehart and Winston

All rights reserved. No part of this publication may be reproduced or transmitted in any form or by any means, electronic or mechanical, including photocopy, recording, or any information storage and retrieval system, without permission in writing from the publisher.

Requests for permission to make copies of any part of the work should be mailed to the following address: Permissions Department, Holt, Rinehart and Winston, 1120 South Capital of Texas Highway, Austin, Texas 78746-6487.

Art and Photo Credits
All work, unless otherwise noted, contributed by Holt, Rinehart and Winston.
Abbreviated as follows: (t) top; (b) bottom; (l) left; (r) right; (c) center; (bkgd) background.
Front cover (owl), Kim Taylor/Bruce Coleman, Inc.; (bridge), Henry K. Kaiser/Leo de Wys; (dove), Stephen Dalton/Photo Researchers, Inc.

Printed in the United States of America

ISBN 0-03-055409-8   3 4 5 6 862 04 03 02

CONTENTS

Chapter 1: The World of Physical Science 1

Chapter 2: The Properties of Matter 6

Chapter 3: States of Matter 9

Chapter 4: Elements, Compounds, and Mixtures 12

Chapter 5: Matter in Motion. 16

Chapter 6: Forces in Motion 21

Chapter 7: Forces in Fluids 24

Chapter 8: Work and Machines 28

Chapter 9: Energy and Energy Resources. 31

Chapter 10: Heat and Heat Technology 36

Chapter 11: Introduction to Atoms 41

Chapter 12: The Periodic Table 44

Chapter 13: Chemical Bonding 48

Chapter 14: Chemical Reactions. 51

Chapter 15: Chemical Compounds 55

Chapter 16: Atomic Energy. 59

Chapter 17: Introduction to Electricity 62

Chapter 18: Electromagnetism 66

Chapter 19: Electronic Technology 69

Chapter 20: Introduction to Waves 73

Chapter 21: The Nature of Sound 77

Chapter 22: The Nature of Light. 82

Chapter 23: Light and Our World 87

CHAPTER 1 DIRECTED READING WORKSHEET

The World of Physical Science

As you read Chapter 1, which begins on page 4 of your textbook, answer the following questions.

Would You Believe . . . ? (p. 4)

1. How did James Czarnowski get his idea for the penguin boat, *Proteus*? Explain.

   Czarnowski got his idea for the penguin boat by watching penguins swim through water. He decided to create a boat that imitates the way penguins swim.

2. What is unusual about the way that *Proteus* moves through the water?

   Proteus moves through the water like a penguin, using "flippers" to propel itself forward.

What Do You Think? (p. 5)

Answer these questions in your ScienceLog now. Then later, you'll have a chance to revise your answers based on what you've learned.

Investigate! (p. 5)

3. What is the seemingly impossible problem in this activity?

   Sample answer: The impossible problem is to fit myself through an index card.

Section 1: Exploring Physical Science (p. 6)

4. Your reflection on the inside of a spoon is different from your reflection on the outside of a spoon. (True) or False? (Circle one.)

DIRECTED READING WORKSHEETS 1

Chapter 1, continued

That's Science! (p. 6)

5. Which of the following activities are part of "doing science"? (Circle all that apply.)
   a. asking questions (c.) making observations
   b. being curious d. taking an opinion poll

6. Only scientists can do science. True or (False)? (Circle one.)

Matter + Energy → Physical Science (p. 7)

7. Physical science is the study of energy and the stuff that everything is made of. (True) or False? (Circle one.)

8. What do air, a ball, and a cheetah have in common?

   Sample answer: All of them are made of matter.

9. All matter has energy, even if it isn't moving. (True) or False? (Circle one.)

10. What is one question you will answer as you explore physical science?

    Accept any reasonable answer. Sample answer: Why does paper burn, while gold does not?

11. Chemistry and physics are both fields of ___physical science___. Chemists study the different forms of ___matter___ and how they interact. ___Energy___ and how it affects ___matter___ are studied in physics.

Looking at the charts on page 8, identify the field of physical science to which each of the following descriptions belongs by writing *physics* or *chemistry* in the space provided.

12. ___physics___ how a compass works
13. ___chemistry___ why water boils at 100°C
14. ___chemistry___ how chlorine and sodium combine to form table salt
15. ___physics___ why you move to the right when the car you are in turns left
16. ___physics___ why you see a rainbow after a rainstorm

2 HOLT SCIENCE AND TECHNOLOGY

Name _____ Date _____ Class _____

Chapter 1, continued

Physical Science Is All Around You (p. 9)

Choose the term in Column B that best matches the topic of physical science in Column A, and write the corresponding letter in the space provided.

Column A	Column B
c 17. waves, currents, and ocean water chemistry	a. ecology
a 18. the transfer of energy in a food chain	b. botany
g 19. weather patterns	c. oceanography
b 20. how plants use carbon dioxide and water to make food	d. biology
f 21. earthquake waves	e. astronomy
e 22. the motion of galaxies in the universe	f. geology
d 23. how the brain sends electrical impulses throughout the body	g. meteorology

Physical Science in Action (p. 10)

24. Which of the careers on page 10 sounds most interesting to you, and why?

Accept any reasonable answer. Sample answer: Being a chemist sounds

most interesting to me because a chemist gets to mix all kinds of

chemicals together and see what happens.

Review (p. 10)

Now that you've finished Section 1, review what you learned by answering the Review questions in your ScienceLog.

Section 2: Using the Scientific Method (p. 11)

1. Advancements in science are mostly due to luck.
True or (False) (Circle one.)

What Is the Scientific Method? (p. 11)

2. Scientists use the scientific method to _____ questions and
_____ problems by following a series of
_____ .
(answer / solve / steps)

Name _____ Date _____ Class _____

Chapter 1, continued

Ask a Question (p. 12)

3. Scientists usually ask a question before they make any observations. True or (False) (Circle one.)

4. ___Efficiency___ is the measure of how much energy is put out by a system compared to how much energy is supplied to a system.

Form a Hypothesis (p. 14)

5. A hypothesis is
   a. not based upon observations.
   b. usually stated in an "If . . . then . . . " format.
   (c.) a possible answer to a question.
   d. best when it is untestable.

Review (p. 14)

Now that you've finished the first part of Section 2, review what you learned by answering the Review questions in your ScienceLog.

Test the Hypothesis (p. 15)

6. Explain the difference between a control group and an experimental group in a controlled experiment.

The control group and experimental group are exactly the same except for

one variable, which is changed in the experimental group.

7. Scientists collect information called _____ data.

Analyze the Results (p. 16)

8. Analysis of results means ___organizing___ data into tables or graphs and looking for the relationships within the data.

Draw Conclusions (p. 17)

9. Which of the following is NOT true when drawing conclusions?
   a. The results may support the hypothesis.
   b. The results may disprove the hypothesis.
   c. The results may neither support nor disprove the hypothesis.
   (d.) None of the above

Chapter 1, continued

Communicate Results (p. 17)

10. What are three ways that communicating results can help other scientists further scientific research?

   Communicating results allows other scientists to confirm the results of tests already performed or to modify certain tests in order to learn something more specific. Scientists can also study a new problem or question based on previous results.

11. Which of the following statements is ALWAYS true for scientific investigations?
   a. Scientists never have a clear idea of the problem they are trying to solve.
   b. Scientists keep testing the same hypothesis.
   (c.) Scientists take accurate measurements and accurately record data.
   d. Scientists follow the steps of the scientific method in the same order.

Building Scientific Knowledge (p. 18)

12. An idea that is supported by many tests and experiments can become a ____theory____ or a ____law____.

13. Which of the following is NOT true of a scientific theory?
   a. It unifies hypotheses and observations that have been supported by testing.
   b. It can predict an observation you might make in the future.
   c. It can be changed or replaced.
   (d.) It is a simple guess.

Mark each of the following statements *True* or *False*.

14. __False__ You could be arrested if you break a scientific law.
15. __False__ Scientific laws are determined by committee.
16. __False__ Laws tell you *why* something happens, not *what* happens.
17. __True__ A scientific law is a summary of many experimental results and observations.

Chapter 1, continued

18. Scientifically speaking, why do you think Figure 13 illustrates the big bang as a theory, not as a law?

   Sample answer: The big bang theory is a theory because it is one explanation for the formation of the universe. Because no one saw the universe form, we don't know for certain if it formed exactly as the explanation says it did. Thus, it can't be a law.

Review (p. 19)

Now that you've finished Section 2, review what you learned by answering the Review questions in your ScienceLog.

Section 3: Using Models in Physical Science (p. 20)

1. What did the MIT engineers hope to gain from making a model?

   The engineers hoped to gain a greater understanding of boat propulsion by making a model.

What Is a Model? (p. 20)

2. You can represent an ____object____ or ____system____ by using a model.

3. Which of the following are ways to use models in science? (Circle all that apply.)
   (a.) looking at the tiny parts of a microscopic cell on a cell diagram
   (b.) launching a homemade rocket in your backyard
   (c.) observing how the parts of matter fit together without being able to see the tiny particles
   (d.) testing a new design for a building on a computer before spending money on construction

Name _____ Date _____ Class _____

Chapter 1, continued

4. How do you think a model rocket might help you understand a real rocket?

Accept all reasonable answers. Sample answer: The model rocket might

help me understand how a real rocket flies into space.

Models Help You Visualize Information (p. 21)

5. In Figure 15, how could using the spring toy as a model help you understand the behavior of sound waves?

Sample answer: The spring toy behaves a lot like sound waves do. It can

show me how air particles crowd together in parts of a sound wave.

6. When we picture things in our minds, we are creating models.
(True) or False? (Circle one.)

Models Are Just the Right Size (p. 22)

Decide whether a useful model for each of the following would be *larger* or *smaller* than the actual object, and write the appropriate answer in the space provided.

7. Mount Everest   smaller

8. a skyscraper   smaller

9. a computer chip   larger

10. the moon   smaller

Models Build Scientific Knowledge (p. 22)

11. Models can be used as _____ tools _____ to illustrate
_____ theories _____ and _____ conduct _____
investigations.

Name _____ Date _____ Class _____

Chapter 1, continued

12. Place each of the following statements in the correct sequence to explain how engineers could use *Proteus* to develop a new technology. Write the appropriate number in the space provided.

   3   Build a full-sized penguin boat.
   2   Discover what factors affect the model's efficiency.
   1   Conduct tests on the model.

13. Why do you think the model in Figure 19 might be useful for understanding atomic theory?

Sample answer: The model might be useful because it helps us see what

atomic theory explains, even though the parts of real atoms are too small

to observe.

Models Can Save Time and Money (p. 23)

14. How can cyber-crashes like the one in Figure 20 save time and money?

Sample answer: Cyber-crashes save money because they don't destroy real

cars. They save time because many different designs can be quickly and

safely tested.

Review (p. 23)

Now that you've finished Section 3, review what you learned by answering the Review questions in your ScienceLog.

Section 4: Measurement and Safety in Physical Science (p. 24)

1. At one time, systems of measurement were based on objects that varied in size, such as body parts and grains of barley.
(True) or False? (Circle one.)

The International System of Units (p. 24)

2. Scientists can _____ share _____ and _____ compare _____ their data when data are expressed in SI units.

Name _____ Date _____ Class _____

Chapter 1, continued

3. Which unit of measurement is most practical for measuring a house?
  a. centimeters (cm)   **c.** meters (m)
  b. kilometers (km)   d. millimeters (mm)

4. Before you can determine how many lenses will fit into a moving crate, what information do you need?

The two pieces of information needed are the volume of the crate and the volume of each lens.

5. Which of the following is NOT a valid unit of measurement for volume?
  a. milliliter   **c.** microgram
  b. cubic centimeter   d. liter

6. Would you measure the mass of a car in milligrams? Why or why not? If not, which unit would you use?

No; measuring the mass of a car in milligrams would not be practical because the value would be very large. It would be more practical to use kilograms or possibly metric tons.

7. The ___kelvin___ is the SI unit for temperature, but scientists often use degrees Celsius.

Derived Quantities (p. 26)

8. List two kinds of derived quantities.

Area and density are derived quantities.

9. What formula would you use to find out how much carpet it would take to cover the floor of your classroom? Write the equation.

The formula is as follows: area = length × width

Name _____ Date _____ Class _____

Chapter 1, continued

10. How could you calculate the mass per unit of volume of the gear in the right column of page 27?

The mass per unit of volume is the density of the object. To calculate the density, divide the mass of the gear by its volume.

Safety Rules! (p. 27)

Match the safety icon in Column B to the correct description in Column A, and write the corresponding letter in the space provided.

Column A	Column B
_i_ **11.** hand safety	a.
_g_ **12.** sharp object	b.
_f_ **13.** clothing protection	c.
_e_ **14.** chemical safety	d.
_a_ **15.** eye protection	e.
_b_ **16.** electrical safety	f.
_h_ **17.** plant safety	g.
_d_ **18.** heating safety	h.
_c_ **19.** animal safety	i.

Review (p. 27)

Now that you've finished Section 4, review what you learned by answering the Review questions in your ScienceLog.

Name _____ Date _____ Class _____

CHAPTER 2 DIRECTED READING WORKSHEET

The Properties of Matter

As you read Chapter 2, which begins on page 34 of your textbook, answer the following questions.

Imagine . . . (p. 34)

1. What are two ways to tell "fool's gold" from real gold?

   One way to tell fool's gold from real gold is that fool's gold shatters when
   it is hit with a hammer, while gold bends. Another way to tell fool's gold
   from real gold is that fool's gold smokes when it is heated, while real gold
   does not.

What Do You Think? (p. 35)

Answer these questions in your ScienceLog now. Then later, you'll have a chance to revise your answers based on what you've learned.

Investigate! (p. 35)

2. What will you do in this activity?

   In this activity I will determine the identity of an object based on its
   properties.

Section 1: What Is Matter? (p. 36)

3. What do a human, hot soup, and a neon sign have in common?

   They are all made of matter.

Everything Is Made of Matter (p. 36)

4. Anything that has _____ volume _____ and
   _____ mass _____ is called matter.

Name _____ Date _____ Class _____

Chapter 2, continued

Matter Has Volume (p. 36)

Mark each of the following statements *True* or *False*.

5. __True__ An object's volume is the amount of space the object takes up.

6. __True__ Things with volume can't share the same space at the same time.

7. __True__ When you measure a volume of water in a graduated cylinder, you should look at the bottom of the meniscus.

8. __False__ A liquid's volume is usually expressed in grams or milligrams.

9. The volume of solid objects is expressed in _____ cubic _____ units. One milliliter is equal to _____ 1 cm³ _____.

10. What three dimensions do you need to know to find the volume of a rectangular solid object?

    You need to know the object's height, width, and length to find its volume.

11. You can't use a ruler to measure a gas, and you can't pour it into a graduated cylinder. So how do you find its volume?

    You can find the volume of a gas by measuring the volume of its container
    because a gas expands to fill its container.

Matter Has Mass (p. 38)

12. List the following objects in order from the least mass to the greatest mass: an elephant, a hamster, a skyscraper, the moon.

    hamster, elephant, skyscraper, the moon

13. The mass of an object should be constant, yet the mass of the puppy in Figure 5, on page 39, will change over time. Explain.

    Sample answer: The mass of the puppy is not constant because matter is
    slowly being added to the puppy as he grows.

Chapter 2, continued

The Difference Between Mass and Weight (p. 39)

14. Why does all matter experience gravity?
 a. All matter has volume. c. All matter is constant.
 (b.) All matter has mass. d. All matter is stable.

15. Look at Figure 6 on page 39. As two objects get closer together _____, the force of gravity between them increases. (closer together or farther apart)

16. Weight is a measure of the gravitational force exerted on an object. (**True**) or False? (Circle one.)

17. A brick weighs less in space than it does on Earth. Why?
A brick weighs less in space than it does on Earth because the gravitational force exerted on it by the Earth decreases as it gets farther away from the Earth.

18. Why do people tend to confuse the terms *mass* and *weight*? (Circle all that apply.)
 (a.) Both remain constant on Earth.
 (b.) People use the terms interchangeably.
 c. Mass is the same thing as weight.
 d. Mass is also dependent on gravity.

Measuring Mass and Weight (p. 41)

19. The unit for __mass__ is the kilogram. The unit for __weight__ is the newton.

Mass Is a Measure of Inertia (p. 42)

20. How do mass and inertia make it easier to pick up an empty juice bottle than a full juice bottle?
A full juice bottle has more mass than an empty one. More mass means more inertia. Because it has more inertia, a full juice bottle is harder to put into motion.

Review (p. 42)
Now that you've finished Section 1, review what you learned by answering the Review questions in your ScienceLog.

Chapter 2, continued

Section 2: Describing Matter (p. 43)

1. In a game of 20 Questions, the most helpful questions you can ask are about the __properties__ of the object.

Physical Properties (p. 43)

2. Which physical properties is the person in the picture on page 43 asking about?
The person in the illustration is asking about the physical properties of color, odor, and mass (or volume).

Match each physical property in Column B to the correct phrase in Column A, and write the corresponding letter in the appropriate space. Use the table on page 44 to help you.

Column A	Column B
__c__ 3. Sand does not dissolve in water.	a. state
__f__ 4. Gold can be made into gold foil.	b. thermal conductivity
__a__ 5. Ice is the solid form of water.	c. solubility
__e__ 6. Copper can be drawn out into wire.	d. density
__b__ 7. A foam cup protects your hand from being burned by the hot chocolate the cup contains.	e. ductility
__d__ 8. Ice cubes float in a glass of water.	f. malleability

9. In the formula for density, D means __density__, V stands for __volume__, and m stands for __mass__.

10. What are two reasons why density is a useful property for identifying substances?
Density is a useful property for identifying substances because a substance's density is always the same at a given temperature and pressure and because most substances have different densities.

ANSWER KEY

Chapter 2, continued

11. Using the table on page 45, list the elements mercury, water, gold, and oxygen from the densest to the least dense.

    gold, mercury, water, oxygen

12. What will happen if you shake the jar in Figure 12? Explain.

    Sample answer: If I shake the jar, the liquids might mix for a minute, but then they'll separate into layers. They separate into layers because the liquids have different densities. The green (densest) liquid will settle to the bottom.

13. Is 40 mL of oil denser than 25 mL of vinegar? _____no_____

14. Density is dependent on the amount of substance you have. True or (False)? (Circle one.)

Review (p. 46)

Now that you've finished the first part of Section 2, review what you've learned by answering the Review questions in your ScienceLog.

Chemical Properties (p. 47)

15. Which of the following are true of chemical properties? (Circle all that apply.)
    a. They indicate a substance's ability to change identity.
    b. They indicate one substance's ability to react with another.
    c. They describe matter.
    d. They can be observed with your senses.

16. In Figure 13, on page 47, why isn't there rust in the painted areas of the car?

    There isn't rust in the painted areas of the car because the paint provides a barrier between oxygen and the steel, which prevents oxygen from reacting with the iron in the steel to form rust.

Chapter 2, continued

Physical vs. Chemical Properties (p. 48)

17. Characteristic properties help scientists to distinguish one substance from another. (True) or False? (Circle one.)

18. Which of the following represents ONLY physical properties? The table on page 48 may help you.
    a. flammable, dense, malleable
    b. malleable, reactive, dense
    c. powdery, dense, red
    d. clear, grainy, nonflammable

Physical Changes Don't Form New Substances (p. 48)

19. What is a physical change?

    A physical change is a change that affects one or more physical properties of a substance.

20. If you make a physical change to a substance, the identity of the substance changes. True or (False)? (Circle one.)

21. Many physical changes are _____easy_____ to undo. (easy or difficult)

Chemical Changes Form New Substances (p. 49)

22. A chemical ___property___ describes a substance's ability to change. A chemical ___change___ occurs when a substance turns into another substance.

23. How do you know that baking a cake involves chemical changes?

    Baking a cake involves chemical changes because the cake has completely different properties than its original ingredients.

24. In Examples of Chemical Changes, the odor of sour milk indicates a chemical change has taken place. (True) or False? (Circle one.)

25. Some chemical changes can be reversed with more chemical changes. (True) or False? (Circle one.)

Review (p. 51)

Now that you've finished Section 2, review what you learned by answering the Review questions in your ScienceLog.

Name _____ Date _____ Class _____

CHAPTER 3 **DIRECTED READING WORKSHEET**

States of Matter

As you read Chapter 3, which begins on page 58 of your textbook, answer the following questions.

Imagine . . . . (p. 58)

1. Lightning sometimes leaves behind a strange calling card. What is it called and how is it formed?

   Lightning strikes silica in the sand and melts it into a liquid. Then the silica

   cools and forms glass. The object formed is called a fulgurite.

2. How do glassmakers use a change of state to make light bulbs, windows, and bottles?

   Glassmakers control the change of state by first melting the silica into liquid

   glass and then shaping the glass before it hardens.

What Do You Think? (p. 59)

Answer these questions in your ScienceLog now. Then later, you'll have a chance to revise your answers based on what you've learned.

Investigate! (p. 59)

3. The purpose of this activity is
   a. to disinfect your hand with alcohol.
   b. to observe a change of state of alcohol.
   c. to observe a change of state of water.

Section 1: Four States of Matter (p. 60)

4. Look at Figure 1. Which of the following states of matter does Hero's steam engine demonstrate? (Circle all that apply.)
   a. solid   **c.** gas
   b. liquid   **d.** plasma

Moving Particles Make Up All Matter (p. 60)

5. The speed of the particles and the strength of the attraction between them determine the ____state of matter____ of the substance.

DIRECTED READING WORKSHEETS **17**

Name _____ Date _____ Class _____

Chapter 3, continued

Match the state of matter in Column A with the description in Column B, and write the corresponding letter in the appropriate space.

Column A	Column B
_a_ 6. Particles have a strong attraction to each other.	a. solid
_c_ 7. Particles move independently of each other.	b. liquid
_b_ 8. Particles are able to slide past one another but do not move independently of each other.	c. gas
_a_ 9. Particles vibrate in place.	
_c_ 10. Particles move fast enough to overcome nearly all of the attraction between them.	

Solids Have Definite Shape and Volume (p. 61)

11. The ship in the bottle in Figure 3 is a solid. How can you tell?

    The ship takes up a certain amount of space and does not take the shape

    of the bottle.

12. Particles that are arranged in a repeating pattern of rows form amorphous solids. True or (False? (Circle one.)

Liquids Change Shape but Not Volume (p. 62)

13. How do the particles of a liquid make it possible to pour juice into a glass?

    The particles in the liquid move quickly and slide past each other until

    the liquid takes the shape of the glass.

14. What does Figure 6 show you about the properties of a liquid?

    It shows that even when liquids change shape, they don't change volume.

15. Liquids tend to form in spherical droplets because of ____surface____ tension.

16. Water has a lower ____viscosity____ than honey.

18 HOLT SCIENCE AND TECHNOLOGY

ANSWER KEY

Chapter 3, continued

Gases Change Both Shape and Volume (p. 63)

17. How is it possible for a cylinder of helium to fill 700 balloons?

The cylinder contains helium particles that are forced close together. As helium enters the balloon, the atoms spread out, and the amount of empty space in the gas increases.

Gas Under Pressure (p. 64)

18. The amount of _____ force _____ exerted on a given area is called pressure.

Review (p. 64)

Now that you've finished the first part of Section 1, review what you learned by answering the Review questions in your ScienceLog.

Laws Describe Gas Behavior (p. 65)

19. The volume of a gas is always the volume of its container. (True) or False? (Circle one.)

20. Boyle's law states that if you keep the temperature constant for a fixed amount of gas, a decrease in pressure means a(n) _____ increase _____ in the volume of the gas.

21. Weather balloons are only partially inflated before they're released into the atmosphere. Why is that?

As the balloon rises, the pressure of the gas decreases as the volume increases. The balloon would pop if it were completely filled before being released.

22. _____ Charles's law _____ is demonstrated by putting a balloon in the freezer.

23. All of the following remain constant in Figure 11 EXCEPT
    a. the type of piston.   (c.) the volume of the gas.
    b. the amount of gas.   d. the pressure.

Chapter 3, continued

Plasmas (p. 67)

Mark each of the following statements *True* or *False*.

24. __True__ More than 99 percent of the known matter in the universe is in the plasma state.

25. __True__ Plasmas are made up of particles that have broken apart.

26. __False__ Plasmas have a definite shape and volume.

27. __False__ Plasmas and gases conduct electric current.

28. __True__ Plasmas are affected by magnetic fields.

29. Lightning and fire are examples of _____ natural _____ plasmas.

30. The incredible light show in Figure 12, on page 67, is caused by plasma. How?

High-energy plasma collides with particles in the upper atmosphere, causing them to glow.

Review (p. 67)

Now that you've finished Section 1, review what you learned by answering the Review questions in your ScienceLog.

Section 2: Changes of State (p. 68)

1. When a substance changes from one _____ physical _____ form to another, we say the substance has had a change of state. True or (False)? (Circle one.)

2. List the five changes of state.

melting, freezing, vaporization, condensation, and sublimation

Energy and Changes of State (p. 68)

3. The identity of a substance changes during a change of state. True or (False)? (Circle one.)

4. Temperature is the measure of the speed of particles. (True) or False? (Circle one.)

5. Temperature is a transfer of energy. True or (False)? (Circle one.)

6. Which has the most energy?
   (a.) particles in steam   c. particles in ice
   b. particles in liquid water   d. particles in freezing water

Chapter 3, continued

Melting: Solids to Liquids (p. 69)

7. Could you use gallium to make jewelry? Why or why not?

No; gallium's melting point is lower than your body temperature. It would melt in your hand.

8. Melting point is a characteristic property, because it is the same for different amounts of the same substance. (True) or False? (Circle one.)

Freezing: Liquids to Solids (p. 69)

9. A substance's freezing point is the temperature at which it changes from a ___liquid___ to a ___solid___.

10. What happens if energy is added or removed from the ice water in Figure 15?

If energy is added, melting occurs. If energy is removed, freezing occurs.

11. Freezing is considered an exothermic change because ___energy___ is removed from the substance.

Vaporization: Liquids to Gases (p. 70)

Choose the term in Column B that best matches the description in Column A, and write the corresponding letter in the space provided.

	Column A	Column B
d	12. vaporization at the surface of a liquid below its boiling point	a. boiling point
b	13. the change of state from a liquid to a gas	b. vaporization
e	14. vaporization that occurs throughout a liquid	c. steam
c	15. the product of vaporization of liquid water	d. evaporation
a	16. temperature at which a liquid boils	e. boiling

Condensation: Gases to Liquids (p. 71)

Mark each of the following statements *True* or *False*.

17. ___False___ At a given pressure, the condensation point for a substance is the same as its melting point.

18. ___True___ For a substance to change from a gas to a liquid, particles must clump together.

19. ___True___ Condensation is an exothermic change.

Sublimation: Solids Directly to Gases (p. 72)

20. Solid carbon dioxide isn't ice. So why is it called "dry ice"?

It's called dry ice because it doesn't melt. It changes from a solid directly into a gas through sublimation.

21. The change of state from a solid to a ___gas___ is called sublimation. Energy must be added for sublimation to occur, so it is an ___endothermic___ change.

Comparing Changes of State (p. 72)

22. Look at the table on page 72. Which two changes of state occur at the same temperature?
 a. condensation and melting
 b. sublimation and freezing
 (c.) vaporization and condensation
 d. melting and vaporization

Temperature Change Versus Change of State (p. 73)

Mark each of the following statements *True* or *False*. Figure 19 may help you.

23. ___True___ The speed of the particles in a substance changes when the temperature changes.

24. ___False___ The temperature of a substance changes before the change of state is complete.

25. ___True___ Energy must be added to a substance to move its temperature from the melting point to the boiling point.

Review (p. 73)

Now that you've finished Section 2, review what you learned by answering the Review questions in your ScienceLog.

Name _____ Date _____ Class _____

CHAPTER 4 DIRECTED READING WORKSHEET

Elements, Compounds, and Mixtures

As you read Chapter 4, which begins on page 80 of your textbook, answer the following questions.

This Really Happened! (p. 80)

1. How do scientists think that the composition of the *Titanic*'s hull caused the "unsinkable" ship to sink?

   The steel of the hull contained large amounts of sulfur, making the steel very brittle. The hull cracked on impact with the iceberg.

2. Why might it be important to learn about the properties of elements, mixtures, and compounds?

   Sample answer: Knowing about their properties might help prevent future disasters.

What Do You Think? (p. 81)

Answer these questions in your ScienceLog now. Then later, you'll have a chance to revise your answers based on what you've learned.

Investigate! (p. 81)

3. What do you think will happen to the ink of the black marker in this activity?

   Accept any reasonable answer. Sample answer: I think that the ink will separate into different colors.

Name _____ Date _____ Class _____

Chapter 4, continued

Section 1: Elements (p. 82)

4. What physical changes can you make to a substance to determine if it's an element? (Circle all that apply.)
   a. crushing (c.) filtering
   (b.) grinding d. passing electric current

An Element Has Only One Type of Particle (p. 82)

5. A pure substance is a substance that contains only one type of particle. (True) or False? (Circle one.)

6. In Figure 2, what do the skillet and the meteorite have in common?

   Both the skillet and the meteorite contain atoms of elemental iron.

Every Element Has a Unique Set of Properties (p. 83)

7. Characteristic properties of elements do NOT depend on the amount of material present in a sample of the element.
   (True) or False? (Circle one.)

8. Why does a helium-filled balloon float up when it is released?

   A helium-filled balloon will float up when released because helium is less dense than air.

9. Look at the properties listed below. Circle the characteristic properties of elements.

   size (melting point) (density) shape
   mass volume (color) surface area
   (hardness) (flammability) weight (reactivity with acid)

10. Suppose you have a cube of nickel and a cube of cobalt, but you don't know which is which. How could you use the characteristic properties listed in Figure 3 to figure out which cube is nickel and which is cobalt?

    Sample answer: Melt the cubes, and note their melting points. The cube that melts at 1,455°C is nickel; the cube that melts at 1,495°C is cobalt.

Name _____ Date _____ Class _____

Chapter 4, continued

Elements Are Classified by Their Properties (p. 84)

11. What are some common properties that most terriers share?

Sample answer: Terriers are small and they have short hair.

12. Which of the following is a property that nickel, iron, and cobalt DON'T share?
    a. shiny
    b. poor conductivity of electric current
    c. good conductivity of thermal energy
    d. None of the above

13. All elements can be classified as metals, metalloids, or nonmetals. (True) or False? (Circle one.)

Look at the chart on page 85. Match the categories of elements in Column B with the correct properties in Column A, and write the corresponding letter in the appropriate space. Categories may be used more than once.

Column A	Column B
_c_ 14. malleable	a. metalloids
_a_ 15. dull or shiny	b. nonmetals
_b_ 16. poor conductors	c. metals
_b_ 17. tend to be brittle and unmalleable as solids	
_c_ 18. always shiny	
_a_ 19. also called semiconductors	
_b_ 20. graphite in pencils	
_b_ 21. always dull	
_a_ 22. used in computer chips	
_c_ 23. ductile	

Review (p. 85)

Now that you've finished Section 1, review what you learned by answering the Review questions in your ScienceLog.

Name _____ Date _____ Class _____

Chapter 4, continued

Section 2: Compounds (p. 86)

1. When two or more elements are chemically combined to form a new pure substance, we call that new substance a _compound_.

2. A compound is different from the elements that reacted to form it. (True) or False? (Circle one.)

3. List three examples of compounds you encounter every day.

Sample answer: table salt, water, and sugar

Elements Combine in a Definite Ratio to Form a Compound (p. 86)

4. Which of the following is NOT true about compounds?
   a. Compounds join in specific ratios according to their masses.
   b. Mass ratios can be written as a ratio or a fraction.
   c. Compounds are random combinations of elements.
   d. Different mass ratios mean different compounds.

Every Compound Has a Unique Set of Properties (p. 87)

Mark each of the following statements *True* or *False*.

5. _True_ Each compound has its own physical properties.
6. _False_ Compounds cannot be identified by their chemical properties.
7. _False_ A compound has the same properties as the elements that form it.
8. Sodium and chlorine can be extremely dangerous in their elemental form. So how is it possible that we can eat them in a compound?

Sample answer: A compound has very different properties than the elements that react to form it. Although sodium and chlorine are dangerous individually, they combine to form sodium chloride, which is harmless, dissolves in water, and is safe to eat.

Name _____ Date _____ Class _____

Chapter 4, continued

Section 3: Mixtures (p. 90)

Properties of Mixtures (p. 90)

1. A pizza is not a mixture. True or (False) (Circle one.)

2. When two or more materials combine without reacting with each other, they form a mixture. (True) or False? (Circle one.)

3. How do the granite in Figure 11 and the pizza at the top of the page show you that the identity of a substance doesn't change in a mixture?

   Sample answer: Because it's possible to identify all of the components in the granite and the pizza, the granite and the pizza show that the identity of a substance doesn't change in a mixture.

4. Mixtures are separated through ___physical___ changes.

Look at the pictures on page 91. Match the technique for separating a mixture in Column B with the substances in Column A, and write the corresponding letter in the appropriate space.

Column A	Column B
a 5. crude oil	a. distill the mixture
d 6. aluminum and iron	b. centrifuge the mixture
b 7. parts of the blood	c. filter the mixture
c 8. sulfur and water	d. pass a magnet over the mixture

9. Granite can be pink or black, depending on the ratio of feldspar, mica, and quartz. (True) or False? (Circle one.)

Review (p. 92)

Now that you've finished the first part of Section 3, review what you learned by answering the Review questions in your ScienceLog.

Name _____ Date _____ Class _____

Chapter 4, continued

Compounds Can Be Broken Down into Simpler Substances (p. 88)

9. What compound helps to give carbonated beverages their "fizz"?
   Carbonic acid helps to give carbonated beverages their "fizz."

   Which elements make up this compound?
   The carbonic acid is made up of the elements carbon, oxygen, and hydrogen.

10. A physical change is the only way to break down a compound. True or (False) (Circle one.)

11. Look at the Physics Connection. The chemical process used to obtain industrial products, such as hydrogen peroxide, is called ___electrolysis___.

Compounds in Your World (p. 89)

12. Which of the following methods are used by living organisms to obtain nitrogen, an element needed to make proteins? (Circle all that apply.)
    a. Bacteria on the roots of pea plants make compounds from atmospheric hydrogen.
    (b.) Plants use nitrogen compounds in the soil.
    (c.) Animals digest plants or animals that have eaten plants.
    d. Plants take in carbon dioxide to make sugar.

13. What do most fertilizers, food preservatives, and medicines have in common?
    Most fertilizers, food preservatives, and medicines are made from manufactured compounds.

14. The compound ___aluminum oxide___ is broken down to produce the element used in cans, airplanes, and building materials.

Review (p. 89)

Now that you've finished Section 2, review what you learned by answering the Review questions in your ScienceLog.

Chapter 4, continued

Solutions (p. 92)

10. Which of the following is NOT true of solutions?
  a. They contain a dissolved substance called a solute.
  b. They are composed of two or more evenly distributed substances.
  c. They contain a substance called a solvent, in which another substance is dissolved.
  (d.) They appear to be more than one substance.

11. In a gaseous or liquid solution, the volume of solvent is _greater than_ the volume of solute.
(less than or greater than)

12. The solid solution used to build the ship *Titanic* was a(n) _alloy_ called steel.

13. Which of the following is true of particles in solutions?
  a. Particles scatter light.
  b. Particles settle out.
  (c.) Particles can't be filtered out of these mixtures.
  d. Particles are large.

14. Concentration is a measure of the volume of a solution.
True or **(False)**? (Circle one.)

15. What is the difference between a concentrated solution and a dilute solution?
Concentrated solutions contain more solute than do dilute solutions.

16. When a solution is holding all the solute it can hold at a given temperature, we say the solution is _saturated_.

17. Solubility is not dependent on temperature. True or **(False)**? (Circle one.)

18. Solubility is expressed in grams of solute per _100 milliliters_ of solvent.

19. Solubility of *gases* in liquids tends to _decrease_ with an increase in temperature. Solubility of *solids* in liquids tends to _increase_ with an increase in temperature.

Chapter 4, continued

20. What are three ways to make a sugar cube dissolve more quickly in water?
You can heat the solution, mix the solution by stirring or shaking it, or crush the sugar before adding it.

Suspensions (p. 96)

21. Which of the following does NOT describe a suspension?
  (a.) Particles are soluble.
  b. Particles settle out over time.
  c. Particles can be seen.
  d. Particles scatter light.

22. Look at the Biology Connection at the top of the page. Why is blood a suspension?
Red blood cells, white blood cells, and platelets are all suspended in plasma.

Colloids (p. 97)

23. What do gelatin, milk, and stick deodorant have in common?
Gelatin, milk, and stick deodorant are all colloids.

Match the mixtures in Column B to the characteristics in Column A, and write the corresponding letter in the appropriate space. Mixtures may be used more than once.

	Column A	Column B
b	**24.** Particles do not settle out.	a. colloids and suspensions
a	**25.** Particles are larger.	b. colloids and solutions
a	**26.** Particles scatter a beam of light.	
b	**27.** Particles cannot be filtered out.	

28. Look at Figure 18. How can a colloid be dangerous to drivers?
The particles in fog, a colloid, scatter light, making it difficult to see the road.

Review (p. 97)
Now that you've finished Section 3, review what you learned by answering the Review questions in your ScienceLog.

CHAPTER 5

DIRECTED READING WORKSHEET

Matter in Motion

As you read Chapter 5, which begins on page 106 of your textbook, answer the following questions.

Would You Believe . . . ? (p. 106)

1. One reason Native Americans played the game of lacrosse was for fun. What was the other reason?

   Native Americans played the game as a substitute for war.

2. What are the advantages of using a lacrosse stick? (Circle all that apply.)
   a. It makes it possible to throw the ball at speeds over 100 km/h.
   b. It helps the defense players hold back the offense.
   c. It makes it possible to throw the ball over 100 km.
   d. It protects the hand from injury from the high speed ball.

What Do You Think? (p. 107)

Answer these questions in your ScienceLog now. Then later, you'll have a chance to revise your answers based on what you've learned.

Investigate! (p. 107)

3. What do you predict this activity will demonstrate?

   Sample answer: I think this activity will demonstrate the connection
   between domino spacing and the speed at which they fall.

Section 1: Measuring Motion (p. 108)

4. Name something in motion that you can't see moving.

   Accept any reasonable answer. Sample answer: I can't see the Earth
   moving, yet I know it moves (revolves) around the sun.

Observing Motion (p. 108)

5. To determine if an object is in motion, compare its position over time to a _____ reference _____ point.

Chapter 5, continued

6. Buildings, trees, and mountains are all useful reference points. Why?

   They make useful reference points because they don't move.

7. Can a moving object be used as a reference point? Explain.

   Yes; a moving object can be used as a reference point because it can be
   observed in relation to another moving object.

Speed Depends on Distance and Time (p. 109)

Each of the following statements is false. Change the underlined word to make the statement true. Write the new word in the space provided.

8. <u>Motion</u> is the rate at which an object moves.
   Speed

9. How fast an object moves depends on the distance traveled and the <u>road</u> taken to travel that distance.
   time

10. The SI unit for speed is <u>km/h</u>.
    m/s, or meters per second

11. Why is it useful to calculate average speed?
    Objects don't often travel at a constant speed.

12. Write out in words how to calculate average speed.
    Average speed is the total distance traveled divided by the total time taken
    to travel that distance.

13. Look at the Brain Food on p. 109. Suppose a car travels 250 m in 10 seconds. Is its average speed greater than or less than that of a running cheetah?

    Its average speed is less than that of a running cheetah.

Chapter 5, continued

Velocity: Direction Matters (p. 110)

14. Why don't the birds in the riddle end up at the same destination?

They traveled in different directions.

15. Velocity has speed and __direction__.

16. Which of the following does NOT experience a change in velocity?

   a. A motorcyclist driving down a straight street applies the brakes.
   b. While maintaining the same speed and direction, an experimental car switches from gasoline to electric power.
   c. A baseball player running from first base to second base at 10 m/s comes to a stop in 1.5 seconds.
   d. A bus traveling at a constant speed turns a corner.

17. To find the resultant velocity, add velocities that are in the __same__ direction(s). Subtract velocities that are in __opposite__ direction(s).

Review (p. 111)

Now that you've finished the first part of Section 1, review what you learned by answering the Review questions in your ScienceLog.

Acceleration: The Rate at Which Velocity Changes (p. 112)

18. Why did the neighbor say you had great acceleration as you slowed down and swerved to avoid hitting a rock?

The neighbor said that because by slowing down and changing direction, I caused a change in my velocity. A change in velocity is acceleration.

19. Write the formula for calculating acceleration in the space below.

$$\text{Acceleration} = \frac{\text{final velocity} - \text{starting velocity}}{\text{time it takes to change velocity}}$$

Chapter 5, continued

20. Scientifically speaking, how do you know the cyclist in Figure 4, on page 113, is accelerating?

Sample answer: A change in velocity means acceleration. Because the cyclist's velocity increased from 1 m/s south to 5 m/s south, I know that the cyclist is accelerating.

21. Another name for acceleration in which velocity increases is __positive__ acceleration.

22. Negative acceleration, or acceleration in which velocity decreases, is also called __deceleration__.

23. What kind of acceleration is occurring in Figure 5, on page 114?

Centripetal acceleration is occurring.

24. When you are standing completely still at the equator, you are accelerating. (True)/ False? (Circle one.)

25. How can you recognize acceleration on a graph?
   a. The graph shows distance versus time.
   b. The graph shows time versus distance.
   c. The graph shows velocity changing as time passes.
   d. The graph is a straight line.

Review (p. 114)

Now that you've finished Section 1, review what you learned by answering the Review questions in your ScienceLog.

Section 2: What Is a Force? (p. 115)

Mark the following statements *True* or *False*.

1. __True__ All forces have size and direction.
2. __True__ A force is a push or a pull.
3. __False__ Forces are expressed in liters.

Name _____ Date _____ Class _____

Chapter 5, continued

Forces Act on Objects (p. 115)

4. You can exert a push without there being an object to receive the push. True or (False)? (Circle one.)

5. Name three examples of objects that you exert forces on when you are doing your schoolwork.

   Accept any reasonable answer. Sample answer: I exert a force on a book when I open it, I exert a force on the keys of a keyboard when I type, and I exert a force on a chair when I sit on it.

6. In which of the following situations is a force being exerted? (Circle all that apply.)
   (a.) A woman pushes the elevator button.
   (b.) A pile of soil sits on the ground.
   (c.) Socks like the ones in Figure 7, on page 116, cling together when they have just come out of the dryer.
   (d.) Magnets stick to the refrigerator.

Forces in Combination (p. 116)

7. In Figure 8, on page 116, how does the net force help the students move the piano?

   When the students apply force in the same direction, their forces are added together to produce a combined net force.

8. Suppose the dog on the left in Figure 9, on page 117, increased its force to 13 N. Which dog would win the tug-of-war? Explain.

   Sample answer: The dog on the left would win the tug-of-war because the net force would be 1 N in favor of the dog on the left.

Name _____ Date _____ Class _____

Chapter 5, continued

Unbalanced and Balanced Forces (p. 117)

9. Why is it useful to know the net force?

   It will help you determine the effect of the force on the motion of an object.

10. Forces are unbalanced when the net force is not equal to zero.

11. To start or change the motion of an object, you need a(n) unbalanced force. (balanced or unbalanced)

12. Forces are balanced when the net force applied to an object is equal to zero. (less than, greater than, or equal to)

13. Are the forces on the cards in Figure 10 balanced? How do you know?

    Yes; if they were unbalanced, the pile of cards would be collapsing.

Review (p. 118)

Now that you've finished Section 2, review what you learned by answering the Review questions in your ScienceLog.

Section 3: Friction: A Force that Opposes Motion (p. 119)

1. What force is responsible for the painful difference between sliding on grass and sliding on pavement?

   The force of friction is responsible.

The Source of Friction (p. 119)

2. Friction occurs when the hills and valleys of two surfaces stick together. (True) or False? (Circle one.)

3. Pavement creates more friction than grass. Why is that?

   Pavement is rougher than grass.

Name _____ Date _____ Class _____

Chapter 5, continued

4. Why is more force needed to slide the larger book in Figure 12, on page 120?

More force is needed because there is more friction between the larger book and the table than between the smaller book and the table.

5. Friction is affected by the amount of surface that is touching.
True or (False)? (Circle one.)

Types of Friction (p. 121)
Match each type of friction in Column B with its example in Column A, and write the corresponding letter in the space provided.

Column A	Column B
a 6. a hockey puck crossing an ice rink	a. sliding friction
d 7. a crate resting on a loading ramp	b. rolling friction
b 8. wheeled cart being pushed	c. fluid friction
c 9. air rushing past a speeding car	d. static friction

10. Static friction is at work if you try to drag a heavy suitcase along the floor and the suitcase _____doesn't move_____.
(moves or doesn't move)

11. As soon as an object starts moving, static friction _____disappears_____. (increases or disappears)

Friction Can Be Harmful or Helpful (p. 123)
12. How does friction harm the engine of a car?
Friction harms the engine of a car by creating heat between moving parts and causing the parts to wear down.

13. Why do you need friction to walk?
Without friction, you would slip and fall when you tried to walk.

DIRECTED READING WORKSHEETS 37

CHAPTER 5

Name _____ Date _____ Class _____

Chapter 5, continued

14. Which of the following are ways to reduce friction? (Circle all that apply.)
(a.) Use a lubricant.
(b.) Make rubbing surfaces smoother.
c. Push surfaces together.
(d.) Change sliding friction to rolling friction.

Review (p. 124)
Now that you've finished Section 3, review what you learned by answering the Review questions in your ScienceLog.

Section 4: Gravity: A Force of Attraction (p. 125)
1. Why did the astronauts in Figure 18 bounce on the moon?
They bounced on the moon because gravity is less on the moon than on the Earth.

2. The force of attraction between two objects due to their masses is the force of _____gravity_____.

All Matter Is Affected by Gravity (p. 125)
3. Does all matter experience gravity? Explain.
Yes; all matter experiences gravity because all matter has mass, and gravity is a result of mass.

4. The force that pulls you toward your pencil is called the _____gravitational_____ force.

5. Look at the Biology Connection at the bottom left of page 125. Scientists think seeds can sense gravity. (True) or False? (Circle one.)

6. Since all objects are attracted to each other due to gravity, why can't you see the objects moving toward each other?
You can't see them moving toward each other because the mass of most objects is too small to cause them to move toward each other.

38 HOLT SCIENCE AND TECHNOLOGY

ANSWER KEY

Chapter 5, continued

7. How are objects around us affected by the mass of the Earth?

Sample answer: Earth has an enormous mass, so its gravitational force is very large. Earth's gravity pulls things toward the Earth's center, and it keeps objects from moving toward each other.

The Law of Universal Gravitation (p. 126)

8. What did Newton figure out about the moon and a falling apple?

He figured out that the force pulling an apple to the Earth was the same unbalanced force that keeps the moon in orbit around the Earth.

9. Newton's law of universal gravitation describes the relationships between all of the following EXCEPT
  a. distance.   c. heat.
  b. mass.   d. gravitational force.

10. Which of the following objects are subject to the law of universal gravitation? (Circle all that apply.)
  a. satellites   c. frogs
  b. water   d. stars

11. If the distance between the objects are the same, the gravitational force between two feathers is greater than the gravitational force between two bowling balls. True or False? (Circle one.)

12. If two objects are moved __toward__ each other, the gravitational force between them increases. (away from or toward)

13. Why is a cat easier to pick up than an elephant?

The cat has less gravitational force acting on it because it has less mass.

14. Read the Astronomy Connection at the bottom right of page 127. In a __black hole__, gravity is so great that even light can't escape.

Chapter 5, continued

15. Why doesn't the sun's gravitational force pull you off the Earth?

The sun is too far away for its gravitational force to have much of an effect on me.

16. What would happen to the Earth and other planets in the solar system without the sun's gravitational force?

Without the sun's gravitational force, planets would not stay in orbit around the sun.

Weight Is a Measure of Gravitational Force (p. 128)

17. The gravitational force exerted by an object depends on the __mass__ of the object. The measure of the Earth's gravitational force on an object is the object's __weight__.

Identify each of the following statements as describing mass or weight. Write M for mass and W for weight.

18. __W__ different on the moon
19. __W__ expressed in newtons
20. __M__ expressed in grams
21. __W__ measure of gravitational force
22. __M__ value doesn't change
23. __M__ amount of matter in an object

24. On Earth, mass and weight are constant, which means they are the same thing. True or False? (Circle one.)

Review (p. 129)

Now that you've finished Section 4, review what you learned by answering the Review questions in your ScienceLog.

ANSWER KEY

CHAPTER 6

DIRECTED READING WORKSHEET

Forces in Motion

As you read Chapter 6, which begins on page 136 of your textbook, answer the following questions.

Imagine ... (p. 136)

1. What is the Vomit Comet?

   The Vomit Comet is the KC-135 airplane, a plane designed to fly at high
   speeds and different angles to simulate the effect of reduced gravity.

2. In this chapter you will learn how __motion__ of objects and how the __gravity__ affects the __laws__ of __motion__ apply to your life.

What Do You Think? (p. 137)

Answer these questions in your ScienceLog now. Then later, you'll have a chance to revise your answers based on what you've learned.

Investigate! (p. 137)

3. What is the purpose of this activity?

   The purpose of this activity is to observe the effect of gravity on a
   falling object.

Section 1: Gravity and Motion (p. 138)

4. Do you agree with what Aristotle might say, that the baseball would land first, then the marble? Explain.

   Accept any reasonable answer. Sample answer: No; I think that the
   baseball would fall faster than the marble because it has a higher
   density.

All Objects Fall with the Same Acceleration (p. 138)

5. Did Galileo prove Aristotle wrong? Explain.

   Yes; Galileo dropped two cannonballs of different masses from the
   Leaning Tower of Pisa. The cannonballs hit at the same time, proving
   that the mass of an object does not affect the rate at which it falls.

6. What does 9.8 m/s/s have to do with acceleration?

   All objects accelerate toward Earth at a rate of 9.8 m/s/s. For every second
   an object falls its downward velocity increases by 9.8 m/s.

Air Resistance Slows Down Acceleration (p. 139)

7. Why does a crumpled piece of paper hit the ground before a flat sheet of paper?

   The flat piece of paper falls more slowly because of air resistance.

8. Air resistance is affected by the __size__ and __shape__ of an object.

9. Air resistance matches the __force of gravity__ when the net force equals zero. (acceleration or force of gravity)

10. When a falling object stops __accelerating__, it has reached __terminal__ velocity.

11. If there were no air resistance, hailstones would
    a. hit the Earth at velocities near 350 m/s.
    b. float gently to the ground like snowflakes.
    c. melt before they hit the ground.
    d. behave exactly as they do now.

12. A sky diver experiences free fall. True or (False) (Circle one.)

13. Free fall occurs because of high air resistance. True or (False) (Circle one.)

Orbiting Objects Are in Free Fall (p. 141)

14. An astronaut is weightless in space. True or (False) (Circle one.)

Name _____ Date _____ Class _____

Chapter 6, continued

15. The shuttle in Figure 7, on page 142, follows the curve of the Earth's surface as it moves forward _____ at a constant speed. At the same time, it is in _____ free fall _____ because of the Earth's gravity.

16. Why don't astronauts hit their head on the ceiling of the falling shuttle?

    Astronauts don't hit their head on the falling shuttle because they are in

    free fall just like the shuttle is. _____

17. Earth's gravity provides a _____ centripetal _____ force that keeps the moon in orbit.

Projectile Motion and Gravity (p. 143)

18. The projectile motion of a leaping frog has two components— _____ vertical _____ and _____ horizontal _____.

Mark each of the following statements *True* or *False*.

19. __False__ The components of projectile motion affect each other.

20. __True__ Horizontal motion of an object is parallel to the ground.

21. __True__ Ignoring air resistance, the horizontal velocity of a thrown object never changes.

22. __True__ On Earth, gravity gives thrown objects their downward vertical motion.

23. If you shoot an arrow aimed directly at the bull's-eye of your target, where will the arrow hit your target? Why?

    It will hit below the bull's eye because the arrow accelerates downward as

    it moves forward. _____

Review (p. 144)

Now that you've finished Section 1, review what you've learned by answering the Review questions in your ScienceLog.

Name _____ Date _____ Class _____

Chapter 6, continued

Section 2: Newton's Laws of Motion (p. 145)

1. In 1686, _____ Sir Isaac Newton _____ published *Principia*, a work explaining _____ three _____ laws to help people understand how forces relate to the _____ motion _____ of objects.

Newton's First Law of Motion (p. 145)

2. What is Newton's first law?

   An object at rest remains at rest and an object in motion remains in

   motion at a constant speed and in a straight line unless acted on by

   an unbalanced force. _____

3. An object in motion would keep moving forever if it never ran into another object or an unbalanced force. (True) or False? (Circle one.)

4. _____ Friction _____ is the unbalanced force that slows down sliding desks, rolling baseballs, and moving cars.

5. How does inertia explain why it would be so difficult to play softball with a bowling ball?

   The bowling ball has more mass than a softball, so it also has more inertia

   than a softball. Having lots of inertia would make it difficult to change the

   bowling ball's direction once it is moving. After the bowling ball is pitched

   toward you, it would be hard for you to bat it away with a bat. _____

Chapter 6, continued

Newton's Second Law of Motion (p. 148)

6. What is Newton's second law of motion?

The acceleration of an object depends on the mass of the object and the

amount of force applied.

7. Look at the Environment Connection in the upper right of page 148. A small car with a small engine cannot accelerate as well as a large car with a large engine. True or (False)? (Circle one.)

8. An object's acceleration decreases as the force on it increases. True or (False)? (Circle one.)

9. Force equals _____mass_____ times _____acceleration_____.

10. The watermelon in Figure 16, on page 149, has more _____mass_____ and _____inertia_____ than the apple, so the watermelon is harder to move than the apple.

Review (p. 149)

Now that you've finished the first part of Section 2, review what you've learned by answering the Review questions in your ScienceLog.

Newton's Third Law of Motion (p. 150)

11. What is Newton's third law of motion?

Whenever one object exerts a force on a second object, the second object

exerts an equal and opposite force on the first. Or, all forces act in pairs.

12. The phrase "equal and opposite" means that the action force and the reaction force have the same _____size_____ but act in opposite _____directions_____.

Chapter 6, continued

13. What action and reaction forces are present when you are sitting on a chair?

The action force is your weight pushing down on the chair. The reaction

force is the force exerted by the chair that pushes up on your body and is

equal to your weight.

14. In a force pair, the reaction and action forces affect the same object. True or (False)? (Circle one.)

15. When a ball falls off a ledge, gravity pulls the ball toward Earth and also pulls Earth toward the ball. (True) or False? (Circle one.)

Momentum Is a Property of Moving Objects (p. 152)

16. Why does it take longer for a large truck to stop than it does for a compact car to stop, even though they are traveling at the same velocity and the same braking force is applied?

It takes the truck longer to stop because it has more momentum.

17. Momentum depends on the _____mass_____ and _____velocity_____ of an object.

18. In Figure 19, on page 152, during the collision, the momentum of the cue ball
   a. is added to the total momentum.
   b. is transferred to the billiard ball.
   c. is transferred to the table holding the balls up.
   d. stays with the cue ball.

19. The law of conservation of momentum states that any time two or more objects interact, they may exchange momentum, but the total amount of momentum stays the same. (True) or False? (Circle one.)

Review (p. 153)

Now that you've finished Section 2, review what you've learned by answering the Review questions in your ScienceLog.

CHAPTER 7

DIRECTED READING WORKSHEET

Forces in Fluids

As you read Chapter 7, which begins on page 160 of your textbook, answer the following questions.

Imagine . . . (p. 160)

1. The ___Mariana Trench___ is the only place on Earth deep enough to swallow the tallest mountain in the world.

2. How might the design of *Deep Flight* allow it to go to such depths and move quickly through the ocean?

   Deep Flight's special hull can withstand the high pressure of deep-sea
   diving. It "flies" through the water because the curvature of its wings
   can be changed.

What Do You Think? (p. 161)

Answer these questions in your ScienceLog now. Then later, you'll have a chance to revise your answers based on what you've learned.

Investigate! (p. 161)

3. How do you think this activity would demonstrate the effect of pressure on a fluid?

   I think I would see a difference in how the water spurts out of holes at
   different heights.

Section 1: Fluids and Pressure (p. 162)

4. How are dogs, flies, dolphins, and humans connected by fluids?

   During their entire lives they move through and breathe fluids.

5. What else can a fluid do besides flow?

   Fluids can take the shape of their container.

---

6. What can particles in a fluid do that particles in a solid can't do?
   a. They can stay rigidly in place.
   b. They can melt.
   (c.) They can move easily past each other.
   d. None of the above

All Fluids Exert Pressure (p. 162)

7. Why does a tire expand as you pump air into it?

   Air is made of particles. More air in the tire means more particles pushing
   against the inside of the tire. More particles pushing means more pressure.

8. To calculate pressure, _____ divide _____ the force exerted by a fluid by the _____ area _____ over which the force is exerted. (multiply or divide, area or volume)

9. Can you blow a bubble that is cube-shaped? Explain.

   No; the air inside the bubble is a fluid. Fluids exert equal pressure in all
   directions, so the bubble expands equally in all directions. Therefore, it
   creates a sphere.

Atmospheric Pressure (p. 163)

Choose the item in Column B that best matches the phrase in Column A, and write your answer in the space provided.

	Column A	Column B
_b_	10. pressure caused by the weight of the atmosphere	**a.** 10
_c_	11. percentage of gases found within 10 km of Earth's surface	**b.** atmospheric pressure
_d_	12. holds the atmosphere in place	**c.** 80
_a_	13. number of newtons pressing on every square centimeter of your body	**d.** gravity

ANSWER KEY

Chapter 7, continued

14. The depth of a fluid is related to the pressure it exerts. The deeper you go in a fluid, the **greater** the pressure becomes. (lower or greater)

15. Look at Figure 4 and number the following locations in order from lowest (1) to highest (5) pressure.
    - **3** Mount Everest's peak
    - **4** La Paz, Bolivia
    - **2** Airplane at cruising altitude
    - **5** Malibu Beach
    - **1** 150,000 m above sea level

16. Why do your ears "pop" as atmospheric pressure decreases?
    Small pockets of air behind my eardrums expand as atmospheric pressure decreases. My ears "pop" when air is released because of the pressure change.

Review (p. 164)
Now that you've finished the first part of Section 1, review what you learned by answering the Review questions in your ScienceLog.

Water Pressure (p. 165)

17. **Water** pressure and **atmospheric** pressure are the two kinds of pressure that contribute to the total pressure on an underwater object.

18. Would you feel more pressure 5 m underwater in a pool or 2 m underwater in a lake? Explain.
    I would feel more pressure in the pool. Pressure depends on depth, not on the total amount of water present.

19. Water is denser than air. (**True**) or False? (Circle one.)

Chapter 7, continued

Fluids Flow from High Pressure to Low Pressure (p. 166)

20. How do you use pressure to sip a drink through a straw?
    As I sip, I remove some of the air in the straw, causing the pressure in the straw to drop. The greater pressure outside the straw pushes on the liquid and forces it up through the straw.

Read each of the following statements, and describe what happens under the conditions described.

21. Pressure is lower inside the lungs than outside the lungs.
    Air flows into the lungs.

22. Pressure is higher inside a tube of toothpaste than outside the tube.
    Toothpaste flows out of the tube.

23. Pressure is higher inside a soda can than outside the can.
    Soda and air spray out of the can.

Pascal's Principle (p. 167)

24. How does a pumping station affect your shower?
    If water pressure is increased at the pumping station, that pressure is transmitted to the rest of the water in the system, all the way to my home. That means I receive good water pressure for my shower.

25. Liquids transmit pressure **more** efficiently than gases do because liquids compress **less** easily than gases. (more or less, more or less)

Name _____ Date _____ Class _____

Chapter 7, continued

26. A hydraulic brake system in a car acts as a force multiplier because the pistons that push the brake pads are _____ larger _____ than the piston that is pushed by the brake pedal. (larger or smaller)

Review (p. 167)

Now that you've finished Section 1, review what you learned by answering the Review questions in your ScienceLog.

Section 2: Buoyant Force (p. 168)

1. How does the buoyant force cause the rubber duck to float?

The buoyant force in the water pushes the rubber duck upward to the

_____ surface of the water. _____

Buoyant Force Is Caused by Differences in Fluid Pressure (p. 168)

2. Buoyant force exists because the pressure in a fluid is greater at the _____ bottom _____ of an object. (top or bottom)

3. What did Archimedes figure out about the buoyant force of an object?

He figured out that the buoyant force is equal to the weight of the

_____ volume of fluid that the object displaces. _____

4. Buoyant force is determined by
- **(a.)** the weight of the displaced water.
- **b.** the volume of the displaced water.
- **c.** the weight of the object.
- **d.** the volume of the object.

Weight vs. Buoyant Force (p. 169)

5. If the weight of the water an object displaces is equal to the weight of the object, the object _____ floats _____. (floats or sinks)

6. If the weight of the water an object displaces is less than the weight of the object, the object _____ sinks _____. (floats or sinks)

Name _____ Date _____ Class _____

Chapter 7, continued

Choose the statement in Column A that best matches the description in Column B, and write the corresponding letter in the space provided.

	Column A	Column B
a	**7.** A rock sinks to the bottom of a pond.	**a.** buoyant force < weight
c	**8.** A duck is buoyed to the surface of a pond.	**b.** buoyant force = weight
b	**9.** A fish floats in a pond.	**c.** buoyant force > weight

An Object Will Float or Sink Based on Its Density (p. 170)

10. How does the density of a rock affect its ability to float?

A rock is denser than water, so it sinks.

11. Why don't most substances float in air if they float in water?

The substances are denser than air but less dense than water.

12. The helium balloon in Figure 11 floats in air because the displaced air is _____ heavier _____ than the helium. (lighter or heavier)

The Mystery of Floating Steel (p. 171)

13. How does the shape of a steel ship allow the ship to float?

The hollow shape increases the volume of the ship, decreasing the ship's

overall density. The overall density of the ship is less than the density of

the water, so the ship floats.

14. When you flood a submarine's tanks with sea water
- **a.** the submarine becomes less dense and rises.
- **b.** the submarine becomes less dense and sinks.
- **c.** the submarine becomes denser and rises.
- **(d.)** the submarine becomes denser and sinks.

ANSWER KEY

Chapter 7, continued

Section 3: Bernoulli's Principle (p. 172)

1. What happens to your shower curtain when you increase the water pressure in your shower?

   The curtain swirls around my legs.

Fluid Pressure Decreases as Speed Increases (p. 173)

2. What did Bernoulli say about the speed of a moving fluid?
   a. The faster the speed is, the higher the pressure.
   b. The slower the speed is, the lower the pressure.
   c. The faster the speed is, the lower the pressure.
   d. Speed and pressure are not related.

3. The table-tennis ball in Figure 14 stays in the water stream. Why?

   The air, which has a higher pressure, pushes the ball into the water stream, which has a lower pressure.

It's a Bird! It's a Plane! It's Bernoulli's Principle! (p. 174)

4. Look at Figure 15. The shape of an airplane wing causes the air above the wing to flow __faster__ than the air below it. (slower or faster)

5. The upward force acting on an airplane wing due to air flow is called __lift__ . (buoyancy or lift)

Chapter 7, continued

15. How does a fish's swim bladder cause the fish to move like a submarine?

    As the swim bladder fills with gases, the fish's overall density decreases and the fish rises. As the swim bladder empties of gases, the fish's overall density increases and the fish sinks.

Review (p. 172)

Now that you've finished Section 2, review what you learned by answering the Review questions in your ScienceLog.

Chapter 7, continued

Thrust and Wing Size Determine Lift (p. 175)

6. How does thrust increase lift?

   Greater thrust means greater speed, which means faster air and increased lift.

7. How large must the wings be for each of the following airplane types? (Circle one for each type.)
   a. high-performance jet   (small)   medium   large
   b. glider   small   medium   (large)

8. Which of the methods below do birds use to stay in the air? (Circle all that apply.)
   (a.) They use large wing size to glide on wind currents.
   b. They pull up their legs.
   c. They move their tails up and down.
   (d.) They flap their wings.

Drag Opposes Motion in Fluids (p. 176)

9. In a strong wind, drag is the _____ force _____ that you walk against.

10. __Turbulence__ usually causes drag forces in flight.

11. Wing flaps are parts of a commercial airplane that reduce drag. (True) or False? (Circle one.)

Wings Are Not Always Required (p. 177)

12. Take a moment to examine Figure 19. A baseball with a side spin will curve __toward__ the side where the ball is moving in the same direction as the air flow. (away from or toward)

Review (p. 177)

Now that you've finished Section 3, review what you learned by answering the Review questions in your ScienceLog.

Name _____ Date _____ Class _____

CHAPTER 8 DIRECTED READING WORKSHEET

Work and Machines

As you read Chapter 8, which begins on page 186 of your textbook, answer the following questions.

Would You Believe . . . ? (p. 186)

1. The Egyptians built the Great Pyramid in 30 years. Why is this so amazing?

   They did not have any heavy-duty machines like cranes, bulldozers, or

   trucks. _____

2. Which of the following are simple machines used by the Egyptians to build pyramids?
   a. screws and screwdrivers
   b. inclined planes and levers
   c. plows and axes
   d. wheels and axles

What Do You Think? (p. 187)

Answer these questions in your ScienceLog now. Then later, you'll have a chance to revise your answers based on what you've learned.

Investigate! (p. 187)

3. What is the purpose of this activity?

   The purpose of this activity is to use a simple machine to make a task

   easier. _____

Section 1: Work and Power (p. 188)

The Scientific Meaning of Work (p. 188)

4. In the scientific sense, you are doing work on this page by reading this question. True or **False**? (Circle one.)

5. When you bowl, you are applying a _____ force _____ to the bowling ball that makes it move down the lane.

Name _____ Date _____ Class _____

Chapter 8, continued

6. When you are carrying a heavy suitcase at a constant speed, you are not doing work. Why not?

   Sample answer: The suitcase is not moving in the same direction as the

   force I am applying to it. _____

7. Look at the chart on page 189. Which of the following is having work done on it? (Circle all that apply.)
   a. a grocery bag as you pick it up
   b. a grocery bag as you carry it at a constant speed
   c. a crate as you push it at a constant speed
   d. a backpack as you are walking with a constant speed

Calculating Work (p. 190)

8. Suppose you want to calculate how much work it takes to lift a 160 N barbell. Besides the weight of the barbell, what other information do you need to know?
   a. the shape of the weights
   b. how high the barbell is being lifted
   c. the strength of the person doing the lifting
   d. None of the above

9. In the equation for work, F is the _____ force _____ applied to the object and d is the _____ distance _____ through which the force is applied.

Power—How Fast Work Is Done (p. 191)

10. The _____ watt _____ is the unit used to express power.

11. If you increase power, you are increasing the amount of work done in a given amount of time. **True** or False? (Circle one.)

Review (p. 191)

Now that you've finished Section 1, review what you learned by answering the Review questions in your ScienceLog.

ANSWER KEY

Chapter 8, continued

Section 2: What Is a Machine? (p. 192)

1. The tools used to fix a flat tire are not complicated enough to be called machines. True or (False)? (Circle one.)

Machines—Making Work Easier (p. 192)

2. Which of the following are machines? (Circle all that apply.)
   - **a.** scissors
   - **b.** lug nut
   - **c.** screw
   - **d.** car jack

Mark each of the following statements *True* or *False*.

3. _True_ As you pry the lid off a can of paint, work is done on the screwdriver and on the lid.

4. _False_ While you pry off the lid with the screwdriver, you are not doing any work.

5. _True_ The force you apply to the screwdriver while prying is the input force.

6. _True_ The work output done by the screwdriver works against the forces of weight and friction to open the lid.

7. _False_ The screwdriver helps you open the can because it increases the amount of work you apply to the can.

8. How does using a ramp make lifting a heavy object easier?
   - **a.** The object is moved over a shorter distance.
   - **b.** The ramp increases the amount of work you do.
   - **c.** Less force is needed to move the object over a longer distance.
   - **d.** None of the above

9. When a machine shortens the distance over which a force is exerted, the size of the force must ___increase___. (increase or decrease)

Mechanical Advantage (p. 196)

10. By comparing the mechanical advantage of two machines, you can tell which machine
    - **a.** is bigger.
    - **b.** has a larger output force.
    - **c.** has a larger input force.
    - **d.** makes work easier.

11. Chopsticks allow you to exert force over a longer distance, but the mechanical advantage is
    - **a.** less than one.
    - **b.** equal to one.
    - **c.** greater than one.
    - **d.** impossible to determine.

Mechanical Efficiency (p. 197)

12. The work output of a machine is always less than the work input. Where does the missing work go?
    - **a.** It is used to get the machine started.
    - **b.** It is used to overcome the friction created by using the machine.
    - **c.** It is used to keep the machine running.
    - **d.** None of the above

13. A machine that has no friction to overcome is called
    - **a.** an ideal machine.
    - **b.** a complex machine.
    - **c.** a real machine.
    - **d.** a smooth machine.

Review (p. 197)

Now that you've finished Section 2, review what you learned by answering the Review questions in your ScienceLog.

Section 3: Types of Machines (p. 198)

1. Name the six simple machines.

   The six simple machines are lever, inclined plane, wedge, screw, wheel and axle, and pulley.

Levers (p. 198)

Mark each of the following statements *True* or *False*.

2. _True_ A first-class lever changes the direction of the input force.

3. _False_ The output force of a second class lever is smaller than the input force.

4. _True_ Third-class levers do not increase the input force.

Inclined Planes (p. 200)

5. You must divide the ___length___ of the inclined plane by the ___height___ you are lifting the load in order to calculate mechanical advantage.

6. An inclined plane saves the work required to lift an object. True or (False)? (Circle one.)

Name _____ Date _____ Class _____

Chapter 8, continued

Wedges (p. 201)

7. Which of the following is NOT a wedge?
   a. knife   **c.** chisel
   b. plow   **(d.)** ramp

8. How do you calculate the mechanical advantage of a wedge?

Sample answer: Divide the length of the wedge by its greatest thickness.

Screws (p. 201)

9. Which of the following is NOT a screw?
   a. jar lid   **c.** bolt
   (b.) steering wheel   **d.** All three are screws.

10. Like using an inclined plane, using a screw enables you to apply a _____small_____ force over _____long_____ distance.

Use what you've learned in the first part of Section 3 to answer the following questions. Choose the machine in Column B that best matches the definition in Column A, and write the corresponding letter in the space provided.

Column A	Column B
_b_ **11.** a straight, slanted surface	**a.** screw
_c_ **12.** a bar that pivots at a fixed point	**b.** inclined plane
_d_ **13.** a double inclined plane that moves	**c.** lever
_a_ **14.** an inclined plane wrapped in a spiral	**d.** wedge

Wheel and Axle (p. 202)

15. In a wheel and axle, which is larger: the radius of the wheel or the radius of the axle?

The radius of the wheel is larger.

Review (p. 202)

Now that you've finished the first part of Section 3, review what you learned by answering the Review questions in your ScienceLog.

◀◀ CHAPTER 8

Name _____ Date _____ Class _____

Chapter 8, continued

16. The input force of a wheel and axle is exerted along
   (a.) a circular distance.
   b. a rectangular distance.
   c. an inclined plane.
   d. a spiral.

Pulleys (p. 203)

17. Which of the following is NOT true of pulleys?
   a. A pulley is a grooved wheel that holds a rope or cable.
   b. A movable pulley moves up with the load as it is lifted.
   c. Fixed and movable pulleys working together form a block and tackle.
   (d.) Fixed pulleys increase force.

Compound Machines (p. 204)

18. A compound machine consists of two or more _____simple machines_____.

19. A can opener is a compound machine. Which of the following simple machines are part of a can opener? (Circle all that apply.)
   (a.) lever   **c.** screw
   (b.) wheel and axle   **(d.)** wedge

20. A machine with many moving parts generally has a lower mechanical efficiency than a machine with fewer moving parts.
   (True) or False? (Circle one.)

Review (p. 205)

Now that you've finished Section 3, review what you learned by answering the Review questions in your ScienceLog.

ANSWER KEY

CHAPTER 9

DIRECTED READING WORKSHEET

Energy and Energy Resources

As you read Chapter 9, which begins on page 212 of your textbook, answer the following questions.

Strange but True! (p. 212)

1. What vast treasure-troves have been buried at sea for millions of years?
   a. gold
   b. gas hydrates
   c. salt
   d. sodium bicarbonate

2. Scientists suspect that large areas off the coasts of North Carolina and South Carolina may contain __70__ times the natural gas consumed by the United States in 1 year.

3. What happens when you hold a flame near icy formations of water and methane?

   The methane ignites, making the hydrate look like a burning ice cube.

What Do You Think? (p. 213)

Answer these questions in your ScienceLog now. Then later, you'll have a chance to revise your answers based on what you've learned.

Investigate! (p. 213)

4. What will you find out in this activity?

   I will find out what happens when energy stored in matter is released.

Section 1: What Is Energy? (p. 214)

5. Where do you think energy is being transferred as the tennis game is played?

   Accept all answers. Students will learn more about the transfer of energy as they work through the section.

Energy and Work—Working Together (p. 214)

6. Energy is the __ability__ to do work.

Chapter 9, continued

7. When you hit a tennis ball with a racket, energy is transferred from the racket to the ball. (**True** or False? (Circle one.)

8. Work and energy are both measured in __joules__.

Kinetic Energy Is Energy of Motion (p. 215)

9. Does the tennis player have kinetic energy if she isn't moving? Explain.

   Sample answer: Yes; the tennis player is still breathing, her eyes are blinking, and her heart is beating. So she still has kinetic energy even if she's standing still.

10. In Figure 2, on page 215, swinging a hammer gives the hammer energy to do work. (**True** or False? (Circle one.)

11. Which of the following have kinetic energy? (Circle all that apply.)
    a. a falling raindrop
    b. a rolling bowling ball
    c. a plane in the sky
    d. a parked car

12. Which of the following is NOT true of kinetic energy?
    a. The faster something moves, the more kinetic energy it has.
    b. The lower the mass is, the higher the kinetic energy.
    c. Speed has a greater effect on kinetic energy than mass has.

13. The truck and the red car in Figure 3, on page 215, are traveling at the same speed. So why does the truck have more kinetic energy?

    The truck is more massive, so it has more kinetic energy.

Potential Energy Is Energy of Position (p. 216)

14. Why does a stretched bow have potential energy?

    The bow has potential energy because work was done on it to change its shape.

Name _____ Date _____ Class _____

Chapter 9, continued

15. Take a moment to look at Figure 5, on page 216. Which of the following would have more gravitational potential energy than a diver on a platform? (Circle all that apply.)
    a. a diver with the same mass on a lower platform
    b. a diver with the same mass on a higher platform
    c. a diver with more mass on the same platform
    d. a diver with less mass on the same platform

16. What two measurements do you multiply together to get gravitational potential energy?
    weight and height

Mechanical Energy Sums It All Up (p. 217)

17. The mechanical energy of the juggler's pins in Figure 6 is the total energy of motion and position of the pins. (True) or False? (Circle one.)

18. Potential energy plus gravitational energy equals mechanical energy. True or (False)? (Circle one.)

Review (p. 217)

Now that you've finished the first part of Section 1, review what you learned by answering the Review questions in your ScienceLog.

Forms of Energy (p. 218)

19. List the six forms of energy.
    thermal, chemical, electrical, sound, light, and nuclear

20. The total potential energy of all the particles in an object is known as thermal energy. True or (False)? (Circle one.)

21. In Figure 7, on page 218, the particles in ocean water have less kinetic energy than the particles in steam. Why?
    The particles in ocean water have less kinetic energy than the particles in steam because the particles of steam move around more rapidly than the particles in ocean water.

Name _____ Date _____ Class _____

Chapter 9, continued

Choose the type of energy in Column B that best matches the definition in Column A, and write the corresponding letter in the space provided. The type of energy may be used more than once.

Column A	Column B
c 22. energy produced by vibrations of electrically charged particles	a. chemical
b 23. energy of a compound that changes when its atoms are rearranged to form a new compound	b. electrical
d 24. energy caused by an object's vibrations	c. light
b 25. energy of moving electrons	d. sound
c 26. energy used in radar systems	

27. Nuclear energy can be produced only by splitting the nucleus of an atom. True or (False)? (Circle one.)

28. Where does the sun get its energy to light and heat the Earth?
    Sample answer: When hydrogen nuclei join together to make helium nuclei, the reaction releases a huge amount of energy.

29. The nucleus of an atom can store _____ potential energy. (potential or kinetic)

Review (p. 221)

Now that you've finished Section 1, review what you learned by answering the Review questions in your ScienceLog.

Section 2: Energy Conversions (p. 222)

1. When you are hammering a nail, what is one type of energy conversion that is taking place?
    Answers should include one of the following: kinetic energy to sound energy, kinetic energy to thermal energy, or chemical energy to kinetic energy.

2. An energy conversion can happen between any two forms of energy. (True) or False? (Circle one.)

Chapter 9, continued

From Kinetic to Potential and Back (p. 222)

Take a look at Figure 14, on page 222. Mark each of the following energy conversions $K \rightarrow P$ (kinetic to potential) or $P \rightarrow K$ (potential to kinetic).

3. __K→P__ You jump down, and the trampoline stretches.
4. __P→K__ The trampoline does work on you, and you bounce up.
5. __K→P__ You reach the top of your jump on the trampoline.
6. __P→K__ You are about to hit the trampoline again.
7. In Figure 15, on page 223, the potential energy of the pendulum is the smallest at which point in its swing?
   a. the highest point   c. the slowest point
   b. the lowest point   d. none of the above

Conversions Involving Chemical Energy (p. 223)

8. Why does eating breakfast give you energy to start the day?
   Sample answer: As my body digests food, the bonds holding the particles of food together are broken. Breaking the bonds releases chemical energy and makes it available to my body for daily activities.

9. The energy you get from food originally comes from the sun.
   (**True**) or False? (Circle one.)

10. During ___photosynthesis___, plants convert light energy into chemical energy.

Conversions Involving Electrical Energy (p. 225)

11. Look at Figure 18. Which of the following forms of energy are converted from electrical energy when you turn on a hair dryer? (Circle all that apply.)
    a. nuclear energy   **c.** kinetic energy
    b. sound energy   **d.** thermal energy

12. In a battery, ___chemical___ energy is converted into ___electrical___ energy.

Review (p. 225)
Now that you've finished the first part of Section 2, review what you learned by answering the Review questions in your ScienceLog.

Chapter 9, continued

Energy and Machines (p. 226)

13. How does a machine make work easier? (Circle all that apply.)
    a. by changing the direction of the required force
    b. by changing the size of the required force
    c. by requiring no force
    d. by increasing the amount of energy transferred

14. A nutcracker can transfer more energy to a nut than you transfer to the nutcracker. True or (**False**)? (Circle one.)

15. Which of the following kinetic energy transfers does NOT occur when you ride a bike as in Figure 20 on page 226?
    a. from legs to pedals   c. from chain to back wheel
    b. from pedals to chain   d. from gear wheel to chain

16. An energy conversion makes the telephone a useful machine. Explain.
    Sample answer: The telephone is a useful machine because it converts sound energy into electrical energy and electrical energy into sound energy.

17. As gasoline burns inside an engine, ___chemical___ energy is converted into thermal and kinetic energy. (electrical or chemical)

Why Energy Conversions Are Important (p. 228)

18. Take a look at Figure 22. How could wind help you cook a meal?
    In a wind turbine, kinetic energy from the wind is converted into electrical energy that I can use to operate a stove and cook my food.

19. The more efficient the light bulb is, the more electrical energy is converted into light energy instead of thermal energy. (**True**) or False? (Circle one.)

20. Energy output is always more than energy input. True or (**False**)? (Circle one.)

Review (p. 228)
Now that you've finished Section 2, review what you learned by answering the Review questions in your ScienceLog.

Name _____ Date _____ Class _____

Chapter 9, continued

Section 3: Conservation of Energy (p. 229)

1. A roller-coaster car never returns to its starting height because energy gets lost along the way. True or (False)? (Circle one.)

Where Does the Energy Go? (p. 229)

2. Where does friction oppose motion on a roller-coaster car? (Circle all that apply.)
   a. between the wheels of the car and the track
   b. between the car and the surrounding air
   c. between the car's axles and the wheels
   d. between the car and the passenger

3. The original amount of potential energy of a roller-coaster car is converted into kinetic energy and ____thermal____ energy as the car races down the hill. (electrical or thermal)

4. The potential energy of the car at the top of the second hill of a roller coaster is equal to the original potential energy of the car at the top of the first hill. True or (False)? (Circle one.)

Energy Is Conserved Within a Closed System (p. 230)

5. A roller coaster is involved in a closed system. List its parts. _track, cars, and surrounding air_

6. The law of conservation of energy states that energy can be neither ____created____ nor ____destroyed____.

7. Look at Figure 24 on page 230. Which forms of energy are converted from the electrical energy that enters the light bulb? (Circle all that apply.)
   a. light energy
   b. thermal energy warming the bulb
   c. thermal energy caused by friction in the wire
   d. chemical energy

No Conversion Without Thermal Energy (p. 231)

8. During an energy conversion, energy is rarely converted to thermal energy. True or (False)? (Circle one.)

Name _____ Date _____ Class _____

Chapter 9, continued

9. Why is it impossible to make a perpetual motion machine?
   Sample answer: A perpetual motion machine would run forever without any additional energy. But because of waste thermal energy that results from energy conversions, the only way to keep a machine moving is to keep putting energy into it.

Review (p. 231)

Now that you've finished Section 3, review what you learned by answering the Review questions in your ScienceLog.

Section 4: Energy Resources (p. 232)

1. How do we use energy resources?
   Accept all reasonable answers. Sample answer: We use energy resources to light and warm our homes, play our stereos, and produce food and clothing.

2. The ____sun____ is the energy resource responsible for most other energy resources.

Nonrenewable Resources (p. 232)

3. Which of the following are fossil fuels? (Circle all that apply.)
   a. coal
   b. wood
   c. petroleum
   d. natural gas

4. Fossil fuels are formed from the remains of plants and animals that lived millions of years ago. (True) or False? (Circle one.)

ANSWER KEY

Chapter 9, continued

Renewable Resources (p. 235)

13. Solar energy can be used to run a television. Explain.

Sample answer: Solar cells in the home can be used to convert solar energy into electrical energy. The electrical energy can be used to run a television.

14. Electrical energy generated from falling water is called hydro-gravity. True or (False)? (Circle one.)

15. A wind turbine converts the _kinetic_ energy of the wind into electrical energy. (potential or kinetic)

16. Geothermal energy results from the heating of Earth's _crust_.

17. Which of the following is NOT an example of biomass?
   a. leaves   c. wood
   b. steel   d. cactus

18. Corn can be used to make a cleaner-burning fuel for cars. (True) or False? (Circle one.)

The Two Sides to Energy Resources (p. 237)

Choose the energy resource in Column B that best matches the disadvantage in Column A, and write the corresponding letter in the space provided.

Column A	Column B
_g_ 19. requires large areas of farmland	a. fossil fuels
_b_ 20. produces radioactive waste	b. nuclear
_d_ 21. requires dams that disrupt river ecosystems	c. solar
_c_ 22. expensive for large-scale energy production	d. water
_f_ 23. waste water can damage soil	e. wind
_e_ 24. only practical in windy areas	f. geothermal
_a_ 25. burning produces smog and acid precipitation	g. biomass

Review (p. 237)

Now that you've finished Section 4, review what you learned by answering the Review questions in your ScienceLog.

Chapter 9, continued

5. Explain why fossil fuels are concentrated forms of the sun's energy.

Sample answer: Plants converted the sun's energy into food during photosynthesis. Animals used and stored this energy by eating the plants or by eating animals that ate plants. The plants and animals died and, over time, their remains were changed into fossil fuels.

Use the images on page 233 to answer questions 6–8.

6. Most of the coal supply in the United States is used for heating. True or (False)? (Circle one.)

7. Which of the following is NOT a petroleum product?
   a. rayon clothing   c. petrochemicals
   b. a candle   **d. wool**

8. The cleanest-burning fossil fuel is _natural gas_.

9. Take a moment to study Figure 27 on page 234. Put the following events in the correct sequence for the production of electrical energy from fossil fuels by writing the appropriate number in the space provided.

   _5_ A large magnet spins within a ring of wire coils.
   _3_ Thermal energy converts liquid water to steam.
   _6_ Electric current is generated in the wire coils.
   _2_ Fossil fuels are burned.
   _4_ Steam pushes against the blades of a turbine.
   _7_ Electrical energy is distributed to homes.
   _1_ Water is pumped into a boiler.

10. In nuclear fission, a nucleus of a(n) _radioactive_ element releases energy when it is split into two nuclei.

11. Nuclear energy is considered to be a renewable resource. True or (False)? (Circle one.)

12. One uranium fuel pellet in Figure 28, on page 235, contains the energy equivalent of about one _metric ton_ of coal.

CHAPTER 10 DIRECTED READING WORKSHEET

Heat and Heat Technology

As you read Chapter 10, which begins on page 244 of your textbook, answer the following questions.

Strange but True! (p. 244)

1. How do Earthship windows maximize the sun's radiant energy?

   They are large and face south.

2. What materials are recycled and used for insulation in the walls of an Earthship?

   Aluminum cans and automobile tires are used for insulation.

What Do You Think? (p. 245)

Answer these questions in your ScienceLog now. Then later, you'll have a chance to revise your answers based on what you've learned.

Investigate! (p. 245)

3. Before you do this activity, which material, if any, do you think will feel the warmest?

   Accept any reasonable answer. Sample answer: All of the materials will feel the same.

Section 1: Temperature (p. 246)

4. Usually when you turn on the hot-water knob, the water that comes out of the faucet isn't hot at first. Can you think of another time when something is labeled "hot" or "cold" and really isn't?

   Accept any reasonable answer. Sample answer: In a car, when you first turn on the heater on a cold day, the air that blows out is not warm.

What Is Temperature? (p. 246)

5. Temperature is a measure of the average _____ of the particles in an object.
   a. potential energy
   b. mechanical energy
   c. kinetic energy
   d. size

6. The faster the particles of an object are moving, the __higher__ the temperature of the object.
   (higher or lower)

7. All of the particles of a substance at a certain temperature move at the same speed and in the same direction. True or (False)? (Circle one.)

8. The temperature of a substance depends on how much of the substance you have. True or (False)? (Circle one.)

Measuring Temperature (p. 248)

9. When a substance undergoes thermal expansion (Circle all that apply.)
   a. its particles expand.
   (b.) its particles spread out.
   c. its particles get colder.
   (d.) its volume increases.

10. All substances will expand equally with the same change in temperature. True or (False)? (Circle one.)

11. According to the Brain Food on page 248, the hottest temperature ever recorded was taken in the Libyan desert. It was 58°C in the shade! What is this temperature on the Fahrenheit scale?
    a. 192°F
    b. 89°F
    (c.) 136°F
    d. 104°F

12. At what temperature does water boil? Write your answer using all three of the temperature scales on page 248.

    Water boils at 212°F, 100°C, and 373 K.

13. It is incorrect to say the temperature of an object is 23°K. Why?

    Temperature on the kelvin scale is reported in kelvin, not in degrees kelvin.

ANSWER KEY

Chapter 10, continued

3. If two objects come in contact with each other, and one object is warmer than the other object, what happens?
   a. Both objects get colder.
   b. Both objects get warmer.
   c. Energy is transferred from the colder object to the warmer object.
   d. Energy is transferred from the warmer object to the colder object.

4. If two objects have the same temperature, but the larger object has more moving particles than the smaller object, what do you know about the thermal energy of the two objects?
   a. The larger object has more thermal energy.
   b. The smaller object has more thermal energy.
   c. They have the same thermal energy.
   d. None of the above

5. The temperature of an object _decreases_ as its particles slow down. (decreases or increases)

6. You place a warm bottle of juice in ice water like in Figure 7 on page 252. After the bottle and the ice water have reached thermal equilibrium
   a. the juice is cooler than the water.
   b. the water is cooler than the juice.
   c. the water and the juice have the same temperature.
   d. the water is cooler than it was before.

Conduction, Convection, and Radiation (p. 253)

7. In conduction, when faster-moving particles collide with slower-moving particles, the faster-moving particles _transfer_ kinetic energy to the slower-moving particles.

8. In the table of conductors and insulators on page 254, what do all of the objects named as conductors have in common?
   a. They are all used for cooking.   c. They have the same shape.
   b. They are all made of metal.   d. They have the same size.

9. In a convection current, warmer particles _rise_ because they are _less dense_ than cooler particles. (rise or sink; denser or less dense)

10. You can feel the warmth of a portable heater like the one in Figure 10, on page 255, when you are standing near it because of _radiation_. (conduction or radiation)

Chapter 10, continued

14. What temperature scale do weather reporters in the United States use to tell you how hot it is outside?
    a. Kelvin   c. Celsius
    b. Fahrenheit   d. SI

15. If you are given a temperature in degrees Fahrenheit and asked to convert it to degrees Celsius, which equation would you use?
    a. $°F = \left(\frac{9}{5} \times °C\right) + 32$ **c.** $°C = \frac{5}{9} \times (°F - 32)$
    b. $K = °C + 273$ d. $°C = K - 273$

More About Thermal Expansion (p. 250)

16. Expansion joints are used in bridges in order to prevent the bridge from
    a. swaying.   **c.** buckling.
    b. getting too hot.   d. making noise.

17. The thermostat in your home has a _bimetallic_ strip that coils or uncoils with changes in _temperature_.

18. Figure 4, on page 250, shows how a thermostat works. When room temperature falls below the desired temperature, the strip in your thermostat _coils_ and an electric circuit is _closed_. (coils or uncoils, opened or closed)

Review (p. 250)
Now that you've finished Section 1, review what you learned by answering the Review questions in your ScienceLog.

Section 2: What Is Heat? (p. 251)

1. A stethoscope feels colder than a tongue depressor when it touches your skin. Why?
   The stethoscope feels colder because of the way energy is transferred between the metal and your skin.

Heat Is a Transfer of Energy (p. 251)

2. Under what condition can heat occur between two objects?
   a. The objects must be hot.
   b. The objects must be large.
   c. The objects must be at different temperatures.
   d. The objects must have a lot of energy.

Name _____ Date _____ Class _____

Chapter 10, continued

11. How is the greenhouse effect helpful?
It is helpful in that it keeps the Earth at a moderate temperature.

12. How is the greenhouse effect harmful?
Excessive greenhouse gases in the atmosphere trap too much thermal energy that might make the Earth too warm.

Review (p. 255)
Now that you've finished the first part of Section 2, review what you learned by answering the Review questions in your ScienceLog.

Heat and Temperature Change (p. 256)

13. The cloth strap of a seat belt has a _higher_ specific heat capacity than the metal buckle, so it takes _less_ energy to change the temperature of the metal than it takes to change the temperature of the cloth. (higher or lower, more or less)

14. Look at the Meteorology Connection on page 256. How does the temperature of the ocean affect the temperature of coastal areas?
During colder months, the ocean keeps nearby areas warm because it retains thermal energy. In the warmer months, the ocean keeps inland areas cool because water doesn't heat up as quickly as the land.

15. Thermal energy is not related to the number of particles in an object. True or (False) (Circle one.)

16. When a substance cools down, the value for heat would be a _negative_ number. (positive or negative)

Name _____ Date _____ Class _____

Chapter 10, continued

17. You can find the specific heat capacity of a substance by using a _calorimeter_. To calculate the specific heat capacity of a substance you need to know the amount of energy transferred by the substance to the water, the test substance's _mass_, _temperature_, and the change in its _temperature_.

18. What is the relationship between a calorie and a Calorie?
One Calorie is equal to 1,000 calories.

19. Look at the can in Figure 15, on page 258. When you read the nutritional information on a package of food, the amount of energy contained in the food in one serving is given in
(a.) Calories.   c. teaspoons.
b. calories.   d. joules.

The Differences Between Temperature, Thermal Energy, and Heat (p. 259)

Indicate whether each of the following is a characteristic of heat, temperature, or thermal energy. In the space provided, write H for heat, T for temperature, and TE for thermal energy.

20. _T_   measure of the average kinetic energy of an object's particles

21. _TE_ varies with the mass and temperature of a substance

22. _TE_ total kinetic energy of a substance's particles

23. _T_   does not vary with the mass of a substance

24. _H_   transfer of energy between objects at different temperatures

Review (p. 259)
Now that you've finished Section 2, review what you learned by answering the Review questions in your ScienceLog.

Section 3: Matter and Heat (p. 260)

1. Why does a frozen juice bar melt before you are done eating it?
Radiation from the sun warms the air around the juice bar. The energy is transferred to the molecules of the juice bar, their kinetic energy increases, and the juice bar turns into a liquid.

Name _____ Date _____ Class _____

Chapter 10, continued

States of Matter (p. 260)

Choose the word in Column B that best matches the description in Column A, and write the corresponding letter in the space provided.

Column A	Column B
b 2. Particles can slide past one another.	a. solid
c 3. Particles move independently of one another.	b. liquid
a 4. Particles vibrate in place.	c. gas

Changes of State (p. 261)

Write the change of state described by each of the following in the appropriate space.

5. liquid changes to gas   boiling
6. liquid changes to solid   freezing
7. gas changes to liquid   condensing
8. solid changes to liquid   melting
9. A change of state changes
   a. the chemical properties of a substance.
   b. the melting point of a substance.
   c. the identity of the substance.
   d. None of the above
10. Look at graph on page 261. When you heat liquid water past 100°C, the temperature of the water does not rise above 100°C until all of the water has become steam.
    (True) or False? (Circle one.)
11. According to the Biology Connection in the upper right of page 262, how do scientists determine the number of Calories in food?
    a. They find the volume of the food sample.
    b. They burn a dry sample in a calorimeter.
    c. They dissolve the food sample in water.
    d. They eat a sample of food and see how fast they can run.

Heat and Chemical Changes (p. 262)
12. What occurs in a chemical change?
    In a chemical change, substances change into new substances with different properties.

Name _____ Date _____ Class _____

Chapter 10, continued

13. Energy is always involved when bonds between particles are broken or formed. (True) or False? (Circle one.)

Review (p. 262)
Now that you've finished Section 3, review what you learned by answering the Review questions in your ScienceLog.

Section 4: Heat Technology (p. 263)

1. Besides a heater, name another example of heat technology.
   Sample answers: automobiles, refrigerators, and air conditioners

Heating Systems (p. 263)
2. In a hot-water heating system, what heats the air in the house?
   The air is heated by radiators and circulates in the room by convection currents.
3. In a warm-air heating system, warm air travels from the furnace through the house in _____ ducts _____ and into rooms through _____ vents _____.
4. If your house has no insulation, what must be done to keep your house warm in the winter?
   The heater must be run almost continuously.
5. Look at Figure 21 on page 264. Air is a good insulator.
   (True) or False? (Circle one.)
6. Which of the following is NOT a component of a passive solar heating system?
   a. large windows facing south   **c.** moving parts
   b. thick walls   d. good insulation
7. Where is water heated in an active solar heating system?
   It is heated in the solar collector.

Chapter 10, continued

Heat Engines (p. 266)
8. In the process of combustion
   - (a.) fuel combines with oxygen to produce thermal energy.
   - b. fuel combines with steam to produce thermal energy.
   - c. air combines with water to produce steam.
   - d. air combines with oxygen to produce steam.

9. A steam engine is an _____external_____ combustion engine. The hot gaseous mixture that expands to do work is _____steam_____.
   (external or internal, steam or fuel)

10. The engine in a car is an _____internal_____ combustion engine. (external or internal)

In a car's engine, the pistons make four strokes. Label the following in the order that the strokes are made by writing the appropriate number in the space provided.

11. __2__ the compression stroke
12. __3__ the power stroke
13. __1__ the intake stroke
14. __4__ the exhaust stroke

Cooling Systems (p. 267)
15. Cooling systems transfer _____thermal_____ energy out of a warm area, so that it feels cooler.

16. Thermal energy moves naturally from areas of _____higher_____ temperature to areas of _____lower_____ temperature.

17. A refrigerator is cool inside because it transfers
   - a. thermal energy from the condenser coils to the inside of the refrigerator.
   - b. specific heat from the condenser coils to the inside of the refrigerator.
   - (c.) thermal energy from inside the refrigerator to the condenser coils.
   - d. specific heat from inside the refrigerator to the condenser coils.

Chapter 10, continued

Heat Technology and Thermal Pollution (p. 269)

18. Thermal pollution hurts a river's ecosystem by
   - a. contaminating the water.
   - b. damaging the river's organisms with smog.
   - (c.) heating the river excessively.
   - d. introducing new species to the ecosystem.

19. What can power plants do to reduce thermal pollution?
   Sample answer: Before hot water is returned to a river, the power plant can reduce the temperature of the water, so that it will have less of an impact on the surrounding ecosystem.

20. Take a moment to read the Environment Connection on page 269. Where do heat islands form? Explain.
   Heat islands form in large cities where excessive amounts of thermal energy are added to the environment. The thermal energy comes from automobiles, factories, and homes.

Review (p. 269)
Now that you've finished Section 4, review what you learned by answering the Review questions in your ScienceLog.

ANSWER KEY

CHAPTER 11

DIRECTED READING WORKSHEET

Introduction to Atoms

As you read Chapter 11, which begins on page 278 of your textbook, answer the following questions.

Would You Believe . . . ? (p. 278)

1. What do dinosaurs have in common with atoms?

   Neither dinosaurs nor atoms can be observed firsthand.

2. How did scientists find information that caused them to change their theory about the way *T. rex* walked? (Circle all that apply.)
   - **a.** by studying well-preserved dinosaur tracks
   - **b.** by examining similarities between the skeletons of *T. rex* and an ostrich
   - c. by observing a *T. rex* as it was walking
   - d. by extracting DNA from fossilized mosquitoes

3. Scientists are able to develop theories about dinosaurs and atoms only through __indirect__ evidence. (direct or indirect)

What Do You Think? (p. 279)

Answer these questions in your ScienceLog now. Then later, you'll have a chance to revise your answers based on what you've learned.

Investigate! (p. 279)

4. How do you think rolling marbles in this activity will help you identify the mystery object?

   When I roll the marble, it will bounce back at different angles. The way the marble bounces back will help me determine the shape of the object.

Section 1: Development of the Atomic Theory (p. 280)

5. Atoms are NOT
   - **a.** a relatively new idea to us.
   - b. the building blocks of all matter.
   - c. the smallest particles into which an element can be divided and still be the same substance.
   - d. seen with the scanning tunneling microscope.

6. An explanation that is supported by testing and brings together a broad range of hypotheses and observations is called a __theory__.

Democritus Proposes the Atom (p. 280)

7. The word *atom* comes from a Greek word that means __indivisible__. (invisible or indivisible)

8. Which of the following statements is part of Democritus's theory about atoms?
   - a. Atoms are small, soft particles.
   - b. Atoms are always standing still.
   - **c.** Atoms join together to form different materials.
   - d. None of the above

9. We know that Democritus was right to say that all matter was made up of atoms. So why did people ignore Democritus's ideas for such a long time?

   Aristotle disagreed with Democritus, and Aristotle had more influence on what __people thought__.

Dalton Creates an Atomic Theory Based on Experiments (p. 281)

10. By conducting experiments and making observations, Dalton figured out that elements combine in random proportions because they're made of individual atoms. True or **False**? (Circle one.)

11. Dalton's theory states that atoms cannot be __created__, __divided__, or __destroyed__.

12. Atoms of different elements are exactly alike. True or **False**? (Circle one.)

Name _____ Date _____ Class _____

Chapter 11, continued

13. How did Dalton think atoms formed new substances?

He thought they formed new substances by joining with other atoms.

Thomson Finds Electrons in the Atom (p. 282)

Mark the following statements *True* or *False*.

14. _False_ In 1897, J. J. Thomson made a discovery that proved the first part of Dalton's atomic theory was correct.

15. _True_ Thomson discovered that there were small particles inside the atom.

16. _True_ Thomson found that the electrically charged plates affected the direction of a cathode-ray tube beam.

17. _False_ Thomson knew the beam was made of particles with a positive charge because it was pulled toward a positive charge.

18. When you rub a balloon on your hair, your hair is attracted _____ to the balloon because both the hair and the balloon have become _____ charged _____.

19. The two types of charge are positive and neutral. True or (False)? (Circle one.)

20. Objects with the same charge attract each other. True or (False)? (Circle one.)

21. In Thomson's "plum-pudding" model, electrons are NOT
  a. negatively charged.
  b. present in every type of atom.
  (c.) collected together in the center of the atom.
  d. scattered throughout a blob of positively charged material.

Review (p. 283)

Now that you've finished the first part of Section 1, review what you learned by answering the Review questions in your ScienceLog.

Rutherford Opens an Atomic "Shooting Gallery" (p. 284)

22. Before his experiment, Rutherford expected the particles to deflect to the sides of the gold foil. True or (False) (Circle one.)

Name _____ Date _____ Class _____

Chapter 11, continued

23. Review Figure 6 and read the text on page 285. Figure 6 shows the new atomic model resulting from Rutherford's experiment. Which of the following statements is NOT part of Rutherford's revision of his former teacher's atomic theory?
  a. Atoms are made mostly of empty space.
  b. The nucleus is a dense, charged center of the atom.
  (c.) Lightweight, negative electrons move in the nucleus.
  d. Most of an atom's mass is in the nucleus.

24. The diameter of the nucleus of an atom is __100,000__ times smaller than the diameter of the atom.

Bohr States That Electrons Can Jump Between Levels (p. 286)

25. In Bohr's atomic model, __electrons__ travel in definite paths around the __nucleus__ in specific levels. Each level is a certain __distance__ from the nucleus. Electrons cannot be found between levels, but they can __jump__ from level to level.

26. Bohr's model only predicted some atomic behavior. (True) or False? (Circle one.)

The Modern Theory: Electron Clouds Surround the Nucleus (p. 286)

27. The exact path of a moving electron can now be predicted. True or (False)? (Circle one.)

28. What are electron clouds?

Electron clouds are regions inside atoms where electrons are likely to be found.

Review (p. 286)

Now that you've finished Section 1, review what you learned by answering the Review questions in your ScienceLog.

ANSWER KEY

Chapter 11, continued

12. Neutrons in the atom's nucleus keep two or more protons from moving apart. (**True**/or False? (Circle one.)

13. If you build an atom using two protons, two neutrons, and two electrons, you have built an atom of ___helium___.

14. An element is composed of atoms with the same number of ___protons___. (neutrons or protons)

Are All Atoms of an Element the Same? (p. 290)

15. It is NOT true that isotopes of an element
  a. have the same number of protons but different numbers of neutrons.
  b. are stable when radioactive.
  c. share most of the same chemical properties.
  d. share most of the same physical properties.

Calculating the Mass of an Element (p. 292)

16. The weighted average of the masses of all the naturally occurring isotopes of an element is called ___atomic___ mass.

What Forces Are at Work in Atoms? (p. 293)

Choose the type of force in Column B that best matches the phrase in Column A, and write the corresponding letter in the space provided.

Column A		Column B
_c_	**17.** counteracts the electromagnetic force so protons stay together in the nucleus	**a.** gravity
_a_	**18.** depends on the mass of objects and the distance between them	**b.** electromagnetic force
_d_	**19.** plays a key role in neutrons changing into protons and electrons in unstable atoms	**c.** strong force
_b_	**20.** holds the electrons around the nucleus	**d.** weak force

Review (p. 293)

Now that you've finished Section 2, review what you learned by answering the Review questions in your ScienceLog.

86 HOLT SCIENCE AND TECHNOLOGY

CHAPTER 11

Section 2: The Atom (p. 287)

1. In this section you will learn about the particles inside the atom and the ___forces___ that act on the particles inside the atom.

How Small Is an Atom? (p. 287)

Each of the following statements is false. Change the underlined word to make the statement true. Write the new word in the space provided.

2. A sheet of aluminum foil is about 500 atoms thick. ___50,000___

3. An Olympic medal contains about twenty thousand billion billion atoms of copper and zinc. ___a penny___

What's Inside an Atom? (p. 288)

Choose the term in Column B that best matches the phrase in Column A, and write the appropriate letter in the space provided.

Column A		Column B
_g_	**4.** particle found in the nucleus that has no charge	**a.** electron cloud
_e_	**5.** particle found in the nucleus that is positively charged	**b.** electron
_f_	**6.** particle with an unequal number of protons and electrons	**c.** amu
_b_	**7.** negatively charged particle found outside the nucleus	**d.** nucleus
_a_	**8.** size of this determines the size of the atom	**e.** proton
_d_	**9.** contains most of the mass of an atom	**f.** ion
_c_	**10.** SI unit used for the masses of atomic particles	**g.** neutron

Review (p. 289)

Now that you've finished the first part of Section 2, review what you learned by answering the Review questions in your ScienceLog.

How Do Atoms of Different Elements Differ? (p. 289)

11. The simplest atom is the ___hydrogen___ atom. It has one proton and one electron.

DIRECTED READING WORKSHEETS 85

Name _____ Date _____ Class _____

CHAPTER 12 DIRECTED READING WORKSHEET

The Periodic Table

As you read Chapter 12, which begins on page 300 of your textbook, answer the following questions.

Would You Believe . . . ? (p. 300)

1. Hyraxes are related to elephants, even though they don't look alike. What have scientists similarly discovered about different-looking elements?

   Sample answer: Scientists have discovered that many different-looking elements have common properties.

2. The periodic table is useful for _____ organizing _____ the _____ properties _____ of unknown elements.

What Do You Think? (p. 301)

Answer these questions in your ScienceLog now. Then later, you'll have a chance to revise your answers based on what you've learned.

Investigate! (p. 301)

3. What will you be looking for in this activity?

   Sample answer: I will be looking for a pattern in the arrangement of the teacher's seating chart.

Section 1: Arranging the Elements (p. 302)

4. Why do you think scientists might have been frustrated by the organization of the elements before 1869?

   Accept any reasonable answer. Sample answer: Scientists might have been frustrated because the elements weren't organized and therefore their properties couldn't be predicted.

DIRECTED READING WORKSHEETS 87

Name _____ Date _____ Class _____

Chapter 12, continued

Discovering a Pattern (p. 302)

5. Mendeleev spent a lot of train rides organizing the elements according to their properties. Which arrangement of elements produced a repeating pattern of properties?
   a. by increasing density
   b. by increasing melting point
   c. by increasing shine
   (d.) by increasing atomic mass

6. How are the days of the week periodic?

   The days of the week are periodic because they have a regular, repeating pattern; they repeat in the same order every 7 days.

7. Mendeleev noticed after arranging the elements that similar _____ physical _____ and _____ chemical _____ properties could be observed in every _____ eighth _____ element.

8. Mendeleev was able to predict the properties of elements that no one knew about. How was this possible?

   Mendeleev was able to predict the properties of unknown elements by using the pattern of properties in the periodic table.

Changing the Arrangement (p. 303)

9. A few elements in Mendeleev's table seemed to be mysteriously out of place according to their properties. How did Moseley solve the mystery? (Circle all that apply.)
   (a.) He rearranged the elements by atomic number.
   b. He discovered protons, neutrons, and electrons.
   c. He disproved the periodic law.
   (d.) He determined the number of protons in an atom.

10. The basis of the periodic table is the periodic _____ law _____, which states that the properties of elements are _____ functions _____ of their atomic _____ numbers _____.

88 HOLT SCIENCE AND TECHNOLOGY

Name _____ Date _____ Class _____

Chapter 12, continued

Use the periodic table on pages 304–305 of your text to fill in the answers to the following questions.

11. Which information is NOT included in each square of the periodic table in your text?
   a. atomic number   **(c.)** melting point
   b. chemical symbol   **d.** atomic mass

12. How can you tell that chlorine is a gas at room temperature?
Sample answer: Chemical symbols are color-coded on the periodic table
according to state. The color of the chemical symbol for chlorine is green,
which corresponds to a gas.

13. Rows of elements are called ___periods___, and columns of elements are called ___groups___ or ___families___.

14. Who will approve the names of the newest elements?
   a. the scientist who discovered each element
   (b.) an international committee of scientists
   c. the chemists from a research institute

15. Silicon is a ___metalloid___ (metal, nonmetal, or metalloid).

Finding Your Way Around the Periodic Table (p. 306)

16. The properties of elements determine whether elements are classified as metals, nonmetals, or metalloids. The number of ___electrons___ in the outer ___energy___ level of an atom helps determine which of these three categories an element belongs to.

17. There is a zigzag line on the periodic table. How can it help you?
The zigzag line can help me recognize which elements are metals, which
are nonmetals, and which are metalloids.

DIRECTED READING WORKSHEETS **89**

◄◄ CHAPTER 12

Name _____ Date _____ Class _____

Chapter 12, continued

Use the pictures on pages 306–307 to help you match the category in Column B with the description in Column A, and write the corresponding letter in the space provided. Categories may be used more than once.

Column A	Column B
_a_ **18.** few electrons in the outer energy level	**a.** metals
_c_ **19.** have some properties of the other two categories	**b.** nonmetals
_b_ **20.** brittle and nonmalleable solids	**c.** metalloids
_b_ **21.** complete or almost-complete set of electrons in the outer energy level	
_a_ **22.** conducts heat from a stovetop to your food	
_b_ **23.** can prevent a spark from igniting gasoline in your car	
_c_ **24.** outer energy level containing a shell of electrons that is about half-complete	
_a_ **25.** formed into electrical wires	
_a_ **26.** flattened into sheets of food wrap without shattering	
_c_ **27.** border the zigzag line on the periodic table	

28. Some elements are named after scientists, like Einstein, and places, like California. (**True**) or False? (Circle one.)

29. The chemical symbol Pb comes from the ___Latin___ word *plumbum*, which means ___lead___.

30. What happens as you move from left to right through each period on the periodic table?
   a. Elements change from having properties of nonmetals to having properties of metals.
   b. Elements change from having properties of metalloids to having properties of metals.
   c. Elements change from liquids to gases.
   (d.) None of the above

Review (p. 309)
Now that you've finished Section 1, review what you learned by answering the Review questions in your ScienceLog.

90 HOLT SCIENCE AND TECHNOLOGY

Name _____ Date _____ Class _____

Chapter 12, continued

Section 2: Grouping the Elements (p. 310)

1. Why do elements in a family or group have similar properties?
   a. They have the same atomic mass.
   b. They have the same number of protons in their nuclei.
   c. They have the same number of electrons in their outer energy level.
   d. They have the same number of total electrons.

Groups 1 and 2: Very Reactive Metals (p. 310)

2. The elements in Groups 1 and 2 are very reactive. Explain.

   Sample answer: Elements in Groups 1 and 2 have only one or two electrons in their outer energy level. They give or share these electrons easily with other elements in order to have a complete set of electrons in their outer energy level.

   _____

   _____

3. Which of the following is NOT true of alkali metals?
   a. They can be cut with a knife.
   b. They are usually stored in water.
   c. They are the most reactive of all the metals.
   d. They can easily give away their outer electron.

4. How are the following alkali metal compounds useful?
   a. sodium chloride _____ used to improve the taste of food
   b. sodium hydroxide _____ used to unclog drains
   c. potassium bromide _____ used in photography

5. Alkaline-earth metals have ___two___ electrons in their outer energy level. They are less reactive and more ___dense___ than alkali metals.

6. Calcium is the alkaline-earth metal that makes up a compound that is healthy for your teeth. (**True** or False? (Circle one.)

▶▶ CHAPTER 12

Name _____ Date _____ Class _____

Chapter 12, continued

Groups 3–12: Transition Metals (p. 312)

7. Besides collectively being called transition metals, Groups 3–12 also have individual names. True or (**False**)? (Circle one.)

8. Which of the following characteristics describe transition metals? (Circle all that apply.)
   a. good conductors
   b. more reactive than alkali and alkaline-earth metals
   c. one or two electrons in the outer energy level
   d. denser than elements in Groups 1 and 2

9. Mercury is different from the other transition metals in Figure 7. How?

   Mercury is different because it is in the liquid state at room temperature.

   _____

10. Two rows of ___transition___ metals are placed at the bottom of the periodic table to save space. Elements in the ___first___ row are called lanthanides and are ___shiny___, ___reactive___ metals.

11. Which lanthanide forms a compound that makes you see red on a computer screen like the one in Figure 8?

    A europium compound makes me see red on a computer screen.

12. All actinides are radioactive. (**True** or False? (Circle one.)

13. Which actinide is used in some smoke detectors?

    The actinide americium is used in some smoke detectors.

Review (p. 313)

Now that you've finished the first part of Section 2, review what you learned by answering the Review questions in your ScienceLog.

Chapter 12, continued

Groups 13–16: Groups with Metalloids (p. 314)

14. Look at Figure 9. The most common element of Group 13, aluminum, was once considered so valuable that Napoleon III used it as dinnerware. (True) or False? (Circle one.)

15. What do diamonds, crayons, and proteins have in common?
    They are all composed of the element carbon.

16. Phosphorous, which makes up about 80 percent of the air you breathe, is used in fertilizers. True or (False)? (Circle one.)

17. All substances need the element oxygen to _____ burn _____.

Complete the following section after you finish reading about Groups 13–16. Each of the following statements is false. Change the underlined word to make the statement true. Write the new word in the space provided.

18. Oxygen group elements contain five electrons in the outer level.
    Nitrogen

19. The carbon group contains no nonmetals.
    boron

20. Nitrogen group and boron group elements vary in reactivity.
    carbon

21. Not all carbon group and oxygen group elements are solid at room temperature.
    nitrogen

Groups 17 and 18: Nonmetals Only (p. 316)

22. Which of the following statements is true?
    a. Group 17 elements are the most reactive metals.
    b. Group 18 elements are the least reactive metals.
    (c.) Group 18 elements are the least reactive nonmetals.

23. What does Figure 12 show about the physical properties of halogens?
    It shows that the physical properties of halogens are not the same.

Chapter 12, continued

24. Halogens are very reactive because of the number of electrons in their outer energy level. (True) or False? (Circle one.)

25. What important use do the halogens iodine and chlorine have in common?
    Both are used as disinfectants.

26. Which of the following statements are true of noble gases? (Circle all that apply.)
    (a.) They are colorless and odorless at room temperature.
    b. They normally react with other elements.
    c. They are metals.
    (d.) They have a complete set of electrons in their outer energy level.

27. Take a moment to look at Figure 13. Why do neon signs contain other noble gases besides neon? Give an example.
    Sample answer: Neon signs contain noble gases besides neon because different noble gases produce different colors. For example, to produce the color lavender, argon is used.

Hydrogen Stands Apart (p. 317)

Mark each of the following statements *True* or *False*.

28. __True__ Hydrogen is useful as rocket fuel.
29. __False__ Hydrogen has two electrons in its outer energy level.
30. __True__ Hydrogen is the most abundant element in the universe.
31. __True__ The physical properties of hydrogen are closer to those of nonmetals than to those of metals.
32. __False__ Hydrogen does not react with oxygen.

Review (p. 317)

Now that you've finished Section 2, review what you learned by answering the Review questions in your ScienceLog.

ANSWER KEY

Name _____ Date _____ Class _____

CHAPTER 13 DIRECTED READING WORKSHEET

Chemical Bonding

As you read Chapter 13, which begins on page 326 of your textbook, answer the following questions.

Strange but True! (p. 326)

1. A scientist discovered superglue by accident. What was he trying to develop?

   He was trying to develop a new plastic for the cockpit bubble of a jet plane.

2. What type of bond is the force that holds atoms together?
   a. electrical
   b. chemical
   c. gravitational
   d. static

3. Which of the following are possible uses of superglue? (Circle all that apply.)
   a. attaching aircraft parts
   b. repairing a cracked tooth
   c. fertilizing plants

What Do You Think? (p. 327)

Answer these questions in your ScienceLog now. Then later, you'll have a chance to revise your answers based on what you've learned.

Investigate! (p. 327)

4. What will you be observing in this activity?
   Sample answer: I will be observing changes in the glue after the borax is added.

Section 1: Electrons and Chemical Bonding (p. 328)

5. Every substance in the world can be made out of about 100 elements. True or False? (Circle one.)

Name _____ Date _____ Class _____

Chapter 13, continued

Atoms Combine Through Chemical Bonding (p. 328)

6. Sugar is made from atoms of which of the following elements? (Circle all that apply.)
   a. carbon
   b. nitrogen
   c. hydrogen
   d. oxygen

7. A chemical bond is the _____ force _____ of attraction that holds a pair of atoms together.

Electron Number and Organization (p. 329)

8. In order to make the overall charge of an atom zero, there must be an equal number of negatively charged _____ electrons _____ and positively charged _____ protons _____.

9. Valence electrons are the electrons in an atom's innermost energy level. True or False? (Circle one.)

Look at Figure 3 on page 330. Write the number of valence electrons for each of the following elements:

10. _6_ oxygen
11. _1_ sodium
12. _7_ chlorine
13. _2_ helium

To Bond or Not to Bond (p. 330)

14. Which electrons determine whether or not an atom will form bonds?
    a. the electrons in the nucleus
    b. the electrons in the innermost energy level
    c. the electrons in the outermost energy level
    d. None of the above

15. An atom will not normally form a chemical bond if it has _____ eight _____ valence electrons.

16. Which of the following does NOT describe how atoms can fill their outermost energy level?
    a. by sharing electrons with other atoms
    b. by losing electrons to other atoms
    c. by gaining electrons from other atoms
    d. by gaining kinetic energy from other atoms

Name _____ Date _____ Class _____

Chapter 13, continued

17. Why does a helium atom need only two valence electrons?

A helium atom needs only two valence electrons because they fill the first energy level, which is also helium's outermost energy level.

Review (p. 331)

Now that you've finished Section 1, review what you learned by answering the Review questions in your ScienceLog.

Section 2: Types of Chemical Bonds (p. 332)

1. Three types of chemical bonds are _____ionic_____, _____covalent_____, and _____metallic_____.

Ionic Bonds (p. 332)

2. Describe how two atoms can become ions.

Sample answer: One atom becomes an ion by gaining an electron, and the other becomes an ion by losing an electron.

3. An atom that loses one or more electrons from its outermost energy level becomes a positively charged ion. (True)/or False? (Circle one.)

4. Which of the following elements give up electrons to other atoms? (Circle all that apply.)
  (a.) sodium   c. chlorine
  (b.) aluminum   d. oxygen

5. Why do the elements in Groups 1 and 2 react so easily?

Very little energy is needed to remove electrons from their atoms.

6. Atoms of nonmetals lose one or more protons when they form ionic bonds. True or (False)? (Circle one.)

Name _____ Date _____ Class _____

Chapter 13, continued

7. The names of negative ions that form when atoms gain electrons have the ending _____-ide_____.

8. A large amount of energy is released when atoms of Group 17 elements lose electrons. True or (False)? (Circle one.)

9. Which of the following are common properties of an ionic compound? (Circle all that apply.)
  (a.) Its solid form is a crystal lattice.
  (b.) It contains alternating positive and negative ions.
  c. It is soft and pliable at room temperature.
  d. Its positive and negative ions repel each other.
  e. It has a low melting point.
  (f.) It has a high boiling point.
  (g.) It is neutral.

10. Look at Figure 8 on page 335. What force causes both the formation of ionic bonds and static cling?
  a. the Earth's gravity
  b. the repulsion of like charges
  (c.) the attraction of opposite charges
  d. a magnetic pole

Review (p. 335)

Now that you've finished the first part of Section 2, review what you learned by answering the Review questions in your ScienceLog.

Covalent Bonds (p. 336)

11. Covalent bonds form between atoms that require a large amount of energy in order to lose an electron. (True)/or False? (Circle one.)

12. In a covalent bond, neither atom loses or gains an electron. Instead, one or more electrons are _____shared_____ by the atoms. (shared or created)

13. Look at Figure 11 on page 336. The electrons that are shared by two atoms spend most of their time
  a. near the smallest of the two atoms.
  b. near the largest of the two atoms.
  (c.) between the nuclei of the two atoms.
  d. in the nuclei of the two atoms.

14. A group of atoms held together by covalent bonds is a neutral particle called a _____molecule_____.

Chapter 13, continued

15. Draw the electron-dot diagram for water.

$$H:\overset{..}{\underset{..}{O}}:H$$

16. Draw the electron-dot diagram for krypton.

$$:\overset{..}{\underset{..}{Kr}}:$$

17. In an electron-dot diagram, each dot represents one proton. True or (False)? (Circle one.)

18. Diatomic molecules are the simplest kinds of molecules. They consist of two atoms bonded together. (True) or False? (Circle one.)

19. Give three examples of complex molecules.

Answers may vary. Sample answer: Gasoline, plastic, and proteins are complex molecules.

20. Carbon is known as the building block of life. Which of the following is a property of this important element? (Circle all that apply.)
   - **a.** Each of its atoms needs to make four bonds.
   - **b.** It is found in all proteins.
   - **c.** It can bond with other elements and form long chains.
   - **d.** It is in a water molecule.

Metallic Bonds (p. 339)

21. In a metal, "swimming" protons surround the metal ions. True or (False)? (Circle one.)

22. What are three properties of metals that are a result of metallic bonding?

The ability to conduct electrical energy, the ability to be reshaped, and the ability to bend without breaking are three properties of metals resulting from metallic bonding.

Chapter 13, continued

23. Because ions in a metal can be easily rearranged without breaking the metallic bonds, metals tend to be easily
   - **a.** shattered.
   - **b.** crystallized.
   - **(c.)** reshaped.
   - **d.** broken.

24. Which of the following is NOT a typical property of a metal?
   - **a.** malleability
   - **b.** ductility
   - **c.** conductivity
   - **(d.)** brittleness

25. Besides being valuable in the jewelry industry, gold is special because it can be hammered into a very thin foil. This property is called
   - **(a.)** malleability.
   - **b.** ductility.
   - **c.** conductivity.
   - **d.** brittleness.

Identify each of the following substances as containing mostly ionic, mostly covalent, or mostly metallic bonds. Refer back to the earlier parts of Section 2 as needed. Write *I* for ionic, *C* for covalent, and *M* for metallic.

26. _M_ copper wire
27. _C_ water
28. _I_ table salt
29. _C_ sugar
30. _C_ carbon dioxide
31. _I_ plaster of Paris
32. _M_ aluminum foil
33. _M_ gold jewelry
34. _I_ sea shells

Review (p. 341)

Now that you've finished Section 2, review what you learned by answering the Review questions in your ScienceLog.

CHAPTER 14 DIRECTED READING WORKSHEET

Chemical Reactions

As you read Chapter 14, which begins on page 348 of your textbook, answer the following questions.

Imagine . . . (p. 348)

The moment an air bag-equipped vehicle slams into a wall, a sequence of events rapidly takes place to protect you from hitting the dashboard. Place the events in the proper sequence by writing the appropriate number in the space provided.

1. __2__ A small electric current is sent to the gas generator.
2. __5__ The air bag fills the space between you and the dashboard.
3. __1__ A sensor detects the sudden decrease in speed.
4. __4__ The air bag inflates with gas formed in the gas generator.
5. __3__ Chemicals in the gas generator react, creating a gas.
6. Could you construct an air bag using vinegar and baking soda for the gas generator? Explain.

Sample answer: No; it would inflate, but not quickly enough to prevent the driver or passenger from hitting the windshield or other hard objects in the car.

What Do You Think? (p. 349)

Answer these questions in your ScienceLog now. Then later, you'll have a chance to revise your answers based on what you've learned.

Investigate! (p. 349)

7. What is the purpose of this activity?

The purpose of this activity is to observe a reaction and identify signs that indicate that a reaction is taking place.

Section 1: Forming New Substances (p. 350)

8. The color of leaves that contain chlorophyll is __green__.

9. The red, orange, and yellow colors of leaves are always present but are hidden until the chlorophyll breaks down. (True)/or False? (Circle one.)

Chemical Reactions (p. 350)

10. Chemical reactions do not change the substances involved. True or (False)? (Circle one.)

11. Look at Figure 2 on page 350. What causes bubbles to form in a muffin?

When baking powder mixes with water, a chemical reaction occurs that produces carbon dioxide and causes bubbles to form in the muffin batter.

Choose the clue to a chemical reaction in Column B that best matches the example in Column A, and write the corresponding letter in the space provided.

Column A	Column B
__b__ 12. thermal energy produced by a fire	a. color change
__c__ 13. precipitate	b. energy change
__d__ 14. bubbles	c. solid formation
__a__ 15. white spots caused by bleach	d. gas formation

16. What does a chemical reaction have to do with making and breaking chemical bonds?

In a chemical reaction, the bonds between atoms in molecules break, the atoms rearrange, and new chemical bonds form. The new arrangement of atoms produces new substances.

Chapter 14, continued

Chemical Formulas (p. 352)

17. The subscript in the chemical formula H_2O tells you there are two
    - **a.** atoms of hydrogen in the molecule.
    - b. electrons on the hydrogen atom in the molecule.
    - c. elements in the molecule.
    - d. atoms of oxygen in the molecule.

In the space provided, write the number of atoms of each element in each of the following chemical formulas.

18. O_2 _____ 2 oxygen
19. $C_6H_{12}O_6$ _____ 6 carbon, 12 hydrogen, 6 oxygen
20. H_2O _____ 2 hydrogen, 1 oxygen
21. $Ca(NO_3)_2$ _____ 1 calcium, 2 nitrogen, 6 oxygen
22. Covalent compounds are usually composed of two or more _____ nonmetals _____ .
23. In the chart on page 353, what number does the prefix *hepta-* stand for?
    - a. six
    - b. eight
    - **c.** seven
    - d. five

Write the formula for each of the following covalent compounds.

24. dinitrogen monoxide _____ N_2O
25. carbon dioxide _____ CO_2
26. Ionic compounds are composed of a _____ metal _____ and a _____ nonmetal _____ .
27. The overall charge of an ionic compound is zero. (**True** or False?) (Circle one.)

Write the formula for each of the following ionic compounds.

28. sodium chloride _____ NaCl
29. magnesium chloride _____ $MgCl_2$

Chemical Equations (p. 354)

30. What do musical notation and chemical equations have in common?
    They both use symbols to communicate information in ways that are easy
    to understand by people who can read those symbols, no matter what
    language they speak.

Chapter 14, continued

31. When carbon reacts with oxygen to form carbon dioxide, carbon dioxide is the _____ product _____ . (product or reactant)

32. Look at Figure 9. In a chemical equation, the formulas of the _____ reactants _____ appear before the arrow, and the formulas of the _____ products _____ appear after the arrow.

33. Which of the following are diatomic elements? (Circle all that apply.)
    - **a.** oxygen
    - b. carbon
    - c. neon
    - **d.** bromine
    - **e.** nitrogen
    - **f.** hydrogen

34. If you make a mistake in a chemical formula, you may be describing a very different substance. (**True** or False? (Circle one.)

35. According to the Brain Food on page 355, what is one good reason to use hydrogen gas as a fuel?
    Water is the only product when hydrogen burns. Burning hydrogen causes
    very little pollution.

36. The coefficient 2 in $2CO_2$ means that there are
    - a. two oxygen atoms and one carbon atom present.
    - b. two oxygen atoms present.
    - c. two carbon atoms and two oxygen atoms present.
    - **d.** two carbon dioxide molecules present.

37. After you finish looking at Figure 11 on page 356, balance the following chemical equation: $O_2 + H_2 \rightarrow H_2O$.

    $O_2 + 2H_2 \rightarrow 2H_2O$

38. Chemical equations must be balanced. Why?
    Chemical equations must be balanced because mass is neither created
    nor destroyed in chemical reactions. For each kind of atom in the reaction,
    the number of atoms in the reactants must equal the number of atoms in
    the products.

Name _____ Date _____ Class _____

Chapter 14, continued

39. Antoine Lavoisier's work led to the law of __conservation__ of mass, which states that mass is neither __created__ nor __destroyed__ in chemical or physical changes.

Review (p. 357)
Now that you've finished Section 1, review what you learned by answering the Review questions in your ScienceLog.

Section 2: Types of Chemical Reactions (p. 358)

1. Which of the following is NOT one of the four classifications of chemical reactions discussed in the text?
   a. synthesis   c. single-replacement
   b. decomposition   **(d.)** double-decomposition

Synthesis Reactions (p. 358)
2. In a synthesis reaction, a single compound is formed from two or more substances. (**True**) or False? (Circle one.)

Decomposition Reactions (p. 359)
3. Decomposition reactions are the __reverse__ of synthesis reactions.

Single-Replacement Reactions (p. 359)
4. How is a person who cuts in on a dancing couple like a single-replacement reaction?

In a single-replacement reaction, one element takes the place of another element that is part of a compound, just as a person who cuts in on a dancing couple takes the place of one dance partner.

5. In a single-replacement reaction, a more reactive element can replace a less reactive element from a compound. (**True**) or False? (Circle one.)

Name _____ Date _____ Class _____

Chapter 14, continued

Double-Replacement Reactions (p. 360)
6. __Ions__ in two __compounds__ switch places in a double-replacement reaction.

After reading Section 2, match these reaction types with their correct examples. Choose the type of reaction from Column B that best matches the example in Column A, and write the corresponding letter in the space provided.

	Column A	Column B
c	7. Zinc reacts with hydrochloric acid to form zinc chloride and hydrogen.	a. decomposition
a	8. Water can be broken down to form hydrogen and oxygen.	b. double-replacement
d	9. Magnesium reacts with oxygen in the air to form magnesium oxide.	c. single-replacement
b	10. Sodium fluoride and silver chloride are formed from the reaction of sodium chloride with silver fluoride.	d. synthesis

Review (p. 360)
Now that you've finished Section 2, review what you learned by answering the Review questions in your ScienceLog.

Section 3: Energy and Rates of Chemical Reactions (p. 361)

1. The rate at which a chemical reaction occurs cannot be changed. True or (**False**)? (Circle one.)

Every Reaction Involves Energy (p. 361)
2. Chemical bonds break as they __absorb__ energy. (absorb or release)
3. An exothermic reaction releases energy. (**True**) or False? (Circle one.)
4. Look at Figure 20 on page 361. Give one example of how energy is released during a chemical reaction.

Sample answers: light from light sticks; electrical energy from flashlight batteries; light and thermal energy from a campfire.

Chapter 14, continued

5. In an endothermic reaction, the chemical energy of the ___reactants___ is less than the chemical energy of the ___products___.

6. What is the law of conservation of energy?

The law of conservation of energy states that energy can be neither created nor destroyed.

7. What happens to the energy absorbed by endothermic reactions?

The energy is stored in the products that form.

8. Which of the following statements about chemical reactions is NOT true?
a. Exothermic reactions release energy.
b. Energy can be stored in molecules.
c. The activation energy of a chemical reaction is the amount of energy released.
d. All chemical reactions require some energy to get started.

9. Once an exothermic reaction is started, it continues to supply the activation energy needed for the substances to react.
True or False? (Circle one.)

Factors Affecting Rates of Reactions (p. 363)

10. What are four factors that affect how rapidly a chemical reaction takes place?

Temperature, concentration, surface area, and the presence of a catalyst or an inhibitor affect reaction rates.

11. The light stick in Figure 23, on page 364, glows brighter in hot water because the rate of a reaction ___increases___ as temperature increases. (decreases or **increases**)

12. The concentration of a solution is a measure of the amount of one substance dissolved in another. **True** or False? (Circle one.)

Chapter 14, continued

13. How does increasing concentration increase the rate of reaction? (Circle all that apply.)
a. There are more reactant particles present, so the particles are more likely to collide with each other.
b. The distance between particles is smaller, so the particles are able to collide more frequently.
c. More particles present gives a lowered activation energy.
d. Increasing the concentration of a solution lowers the surface area of the reactants.

14. What is one way you can increase the surface area of a solid reactant?

One way to increase the surface area of a solid reactant is to grind the reactant to a powder.

15. A catalyst lowers the activation energy of a reaction.
True or False? (Circle one.)

16. The catalysts used in your body are called ___enzymes___.

17. Which of the following slows down a reaction?
a. platinum in a catalytic converter
b. a food preservative
c. an enzyme in your body
d. None of the above

18. Which of the following can increase the rate of a chemical reaction? (Circle all that apply.)
a. raising the temperature
b. reducing the amount of one substance dissolved in another
c. increasing the surface area of a solid reactant
d. adding a catalyst
e. adding an inhibitor

Review (p. 365)
Now that you've finished Section 3, review what you learned by answering the Review questions in your ScienceLog.

Name _____ Date _____ Class _____

CHAPTER 15 DIRECTED READING WORKSHEET

Chemical Compounds

As you read Chapter 15, which begins on page 372 of your textbook, answer the following questions:

Strange but True . . . (p. 372)

1. What was the inventor of Silly Putty® trying to do?

   The inventor of Silly Putty was trying to find a replacement substance for
   rubber.

What Do You Think? (p. 373)

Answer these questions in your ScienceLog now. Then later, you'll have a chance to revise your answers based on what you've learned.

Investigate! (p. 373)

2. The purpose of this activity is to see how the type of __bonding__ in compounds can determine the __properties__ of the compounds.

Section 1: Ionic and Covalent Compounds (p. 374)

3. Which of the following is NOT true about chemical compounds?
   a. Chemical compounds are composed of molecules or ions.
   b. There are actually only a few kinds of chemical compounds.
   c. Chemical compounds have different kinds of bonds.
   d. Chemical compounds are all around us.

Ionic Compounds (p. 374)

4. An ionic bond forms when electrons are transferred from metal atoms to nonmetal atoms. (**True** or False? (Circle one.)

5. When sodium reacts with chlorine, __salt or sodium chloride__ forms.

6. A crystal lattice is a __repeating__ three-dimensional pattern of ions.

7. In a crystal lattice, each ion is surrounded by and bonded to ions with the same charge. True or (**False**) (Circle one.)

Name _____ Date _____ Class _____

Chapter 15, continued

8. Circle the properties of ionic compounds.
   (**strong bonds**)   weak bonds
   hard to break   (**shatter easily**)
   (**high melting point**) low melting point
   (**solid at room temperature**) liquid at room temperature
   difficult to dissolve in water (**easy to dissolve in water**)

9. __Undissolved__ ionic compounds cannot conduct an electric current. (**Dissolved** or Undissolved)

Covalent Compounds (p. 375)

10. Which of the following statements is NOT true of covalent compounds?
    a. They form when atoms of two elements share electrons.
    b. They have weaker bonds than ions in a crystal lattice.
    c. They are composed of independent particles called molecules.
    d. They have a higher melting point than ionic compounds.

11. Why don't water and oil mix?
    Sample answer: The molecules of water have a greater attraction to each
    other than to the oil, and they squeeze the oil out of the way.

12. Most solutions that contain molecules of covalent compounds do not conduct an electric current. (**True** or False? (Circle one.)

Review (p. 376)

Now that you've finished Section 1, review what you learned by answering the Review questions in your ScienceLog.

Section 2: Acids, Bases, and Salts (p. 377)

1. Lemon changes the color of tea because the lemon contains a substance called a(n) __acid__. (**acid** or base)

Name _____ Date _____ Class _____

Chapter 15, continued

Acids (p. 377)

2. Why should you never use taste as a test to identify an unknown chemical?

Never use taste as a test because acids can be corrosive or poisonous.

3. Acids react with all metals to produce hydrogen gas. True or (False)? (Circle one.)

4. When an acid is placed in water, the number of hydrogen ions, H^+, **increases**. These extra hydrogen ions combine with **water** molecules to form hydronium (or H_3O^+) ions.

5. A substance that changes color in the presence of an acid or a base is called an indicator. (True) or False? (Circle one.)

6. Blue litmus paper tests for the presence of bases. True or (False)? (Circle one.)

Choose the acid in Column B that best matches each use in Column A, and write the corresponding letter in the space provided. Acids may be used more than once.

Column A		Column B
c	**7.** paper and paint production	**a.** citric acid
e	**8.** the "bite" in soft drinks	**b.** nitric acid
d	**9.** digestion	**c.** sulfuric acid
a	**10.** in orange juice	**d.** hydrochloric acid
d	**11.** algae preventative	**e.** carbonic acid
b	**12.** rubber production	

13. An acid is strong if all of its molecules break apart in water to produce hydrogen ions. (True) or False? (Circle one.)

14. Which of the following are weak acids? (Circle all that apply.)
sulfuric acid   (carbonic acid)
(phosphoric acid)   (citric acid)
nitric acid   hydrochloric acid

Name _____ Date _____ Class _____

Chapter 15, continued

Bases (p. 379)

15. Bases have a **bitter** taste and feel **slippery**.

16. Bases increase the number of hydrogen ions in a solution. True or (False)? (Circle one.)

Match each of the bases in Column B with the common uses in Column A, and write the corresponding letter in the space provided.

Column A		Column B
d	**17.** treating heartburn	**a.** ammonia
c	**18.** unclogging drains and making soap	**b.** calcium hydroxide
b	**19.** making cement	**c.** sodium hydroxide
a	**20.** household cleaning	**d.** magnesium hydroxide

21. A base is strong when all the molecules break apart in water to produce hydrogen ions. True or (False)? (Circle one.)

Acids and Bases Neutralize One Another (p. 380)

22. How do antacids get rid of heartburn?

The base in the antacid neutralizes the acid in your stomach.

23. H^+ ions of an acid and OH^- ions of a base combine to form the compound we know as **water**.

24. A solution with a pH of 3 is **acidic**. (basic or acidic)

25. Which of the following can be used to determine the pH of a solution? (Circle all that apply.)
(**a.**) a mixture of indicators   **c.** litmus paper
(**b.**) a pH meter   **d.** lemon juice

26. Living things need a steady **pH** in their environment.

27. Why is the helicopter in Figure 15, on page 381, dumping a base in the lake?

The base will neutralize the acid in the lake and makes it a better environment for the organisms living there.

ANSWER KEY

Name _____ Date _____ Class _____

Chapter 15, continued

Salts (p. 382)

28. The __positive__ ion of a base and the __negative__ ion of an acid can combine to form a(n) __ionic__ compound called a salt.

29. Salts can be produced in many different reactions. List the three reactions shown in Figure 16 on page 382.
    neutralization of an acid and a base, reaction of a metal with an acid, and reaction of a metal with a nonmetal

Mark each of the following statements *True* or *False*.

30. __False__ Calcium chloride is used to season your food.
31. __True__ Salt can keep the roads ice-free in winter.
32. __True__ The walls of your room may contain calcium sulfate.
33. __True__ Sodium nitrate is used in food preservation.

Review (p. 382)

Now that you've finished Section 2, review what you learned by answering the Review questions in your ScienceLog.

Section 3: Organic Compounds (p. 383)

1. Organic compounds are composed of __carbon-based__ molecules.

Each Carbon Atom Forms Four Bonds (p. 383)

2. Carbon atoms are able to bond with up to four other atoms.
   (**True**) or False? (Circle one.)

3. Take a look at Figure 18 on page 383. Which of the following is NOT a type of carbon backbone?
   a. carbon atoms connected in a line
   b. a chain continuing in more than one direction
   c. a chain forming a ring
   d. None of the above

Biochemicals: The Compounds of Life (p. 384)

4. Biochemicals are organic compounds made by living things.
   (**True**) or False? (Circle one.)

DIRECTED READING WORKSHEETS **113**

CHAPTER 15

Name _____ Date _____ Class _____

Chapter 15, continued

5. What are carbohydrates? (Circle all that apply.)
   (**a.**) biochemicals
   (**b.**) sources of energy
   (**c.**) one or more simple sugars bonded together

Identify each item below as describing simple (S) or complex (C) carbohydrates.

6. __C__ many sugar molecules bonded together
7. __S__ produced by plants through photosynthesis
8. __C__ bread, cereal, and pasta
9. __C__ stored, extra sugar
10. __S__ one or a few sugar molecules bonded together
11. __S__ fruits and honey

12. Which of the following is NOT an example of a lipid?
    a. candle   (**c.**) sugar
    b. chicken fat   d. corn oil

13. Lipids store energy, make up cell membranes, and do not dissolve in water. (**True**) or False? (Circle one.)

14. Lipids store excess __energy__ in the body.
    Plants tend to store lipids as __oils__.
    Animals tend to store lipids as __fats__.

Mark each of the following statements about phospholipid molecules *True* or *False*.

15. __True__ The molecules help control the movement of chemicals into or out of the cell.
16. __False__ The head of a molecule is a long carbon backbone composed of only carbon and hydrogen atoms.
17. __True__ Water is attracted to the heads of the molecules.
18. __False__ The cell membrane is mostly made of three layers of the molecules.

19. Take a look at the Brain Food in the right column of page 385. Do you think we should eliminate cholesterol from our diets? Explain.
    Sample answer: No, because cholesterol is needed in nerve and brain tissue and for making growth hormones.

20. Biochemicals composed of __amino acids__ are called proteins.

114 HOLT SCIENCE AND TECHNOLOGY

Chapter 15, continued

21. Which of the following are functions of proteins? (Circle all that apply.)
   a. regulate chemical activities
   b. store excess sugar
   c. transport and store materials
   d. provide support

22. Proteins are all the same size and shape. True or False? (Circle one.)

23. The shape adopted by the bonded amino acids determines the function of the protein. True or False? (Circle one.)

24. Insulin is one of the smallest proteins. What very important job does this small protein have? Insulin helps regulate your blood sugar level.

25. The protein in your blood that carries oxygen to all parts of your body is called ___hemoglobin___.

26. Some large proteins help control the transport of materials into and out of cells. True or False? (Circle one.)

27. Why are nucleic acids sometimes called the "blueprints of life?" Nucleic acids are called that because they contain all the information that a cell needs to make all of its proteins.

28. Nucleic acids are composed of atoms of carbon, phosphorous, hydrogen, ___nitrogen___, and ___oxygen___.

Mark each of the following statements *True* or *False*.

29. True   The two types of nucleic acids are DNA and RNA.
30. False RNA is the only genetic material in the cell.
31. False DNA molecules can only store a tiny amount of information due to their small size.
32. False RNA contains the information to make DNA.
33. True   RNA is involved in protein-building.

Review (p. 387)
Now that you've finished the first part of Section 3, review what you learned by answering the Review questions in your ScienceLog.

Chapter 15, continued

Hydrocarbons (p. 388)
For each of the following statements, identify the hydrocarbons as saturated (S), unsaturated (U), or aromatic (A).

34. U   At least two carbon atoms share a double or a triple bond.
35. S   Each carbon atom shares a single bond with four other atoms.
36. A   Air fresheners and mothballs contain this type of hydrocarbon.
37. S   No atoms can be added without replacing an atom that is part of the hydrocarbon.
38. A   Most are based on benzene.
39. A   Many medicines are manufactured using this type of hydrocarbon.
40. U   Ethene helps ripen fruit.
41. S   Propane is this type of hydrocarbon.

Other Organic Compounds (p. 389)

42. All organic compounds are made of only carbon and hydrogen. True or False? (Circle one.)

Match each organic compound in Column A with one of its uses in Column B, and write the corresponding letter in the space provided.

	Column A	Column B
c	**43.** food preservative	**a.** ester
a	**44.** fragrance	**b.** alkyl halide
d	**45.** antifreeze	**c.** organic acid
b	**46.** refrigerant	**d.** alcohol

Review (p. 389)
Now that you've finished Section 3, review what you learned by answering the Review questions in your ScienceLog.

CHAPTER 16 DIRECTED READING WORKSHEET

Atomic Energy

As you read Chapter 16, which begins on page 396 of your textbook, answer the following questions.

Would You Believe . . . ? (p. 396)

RTGs are useful in places where replacing a battery or using solar power would be difficult. In items 1–3, explain why using solar power is difficult in each of the following places: Accept all reasonable answers. Samples are given.

1. near Saturn

   Saturn is very far away from the sun.

2. in the deep ocean

   It is very dark in the deep ocean.

3. on the Arctic icecap

   The icecap gets little sunlight in winter.

What Do You Think? (p. 397)

Answer these questions in your ScienceLog now. Then later, you'll have a chance to revise your answers based on what you've learned.

Investigate! (p. 397)

4. What is the purpose of this activity?

   The purpose of this activity is to model the decay of unstable nuclei into stable nuclei.

Section 1: Radioactivity (p. 398)

5. Almost everyone has learned something by obtaining unexpected results. Discuss a time when this happened to you or one of your friends.

   Accept any reasonable answer. Sample answer: When I put a can of soda in the freezer, I did not expect it to expand. I learned that water expands when it freezes.

Discovering Radioactivity (p. 398)

6. Items that glow when exposed to light are made from __fluorescent__ materials.

7. Why did Becquerel place paper over the photographic plate?
   a. so he could protect the plate from scratches
   b. so he could protect the fluorescent mineral from scratches
   c. so light would not reach the plate
   d. so the X rays would not reach the plate

8. __Uranium__ was the element in the fluorescent mineral that gave off the invisible energy.

9. Marie Curie used the term *radioactivity* to describe what property?

   Marie Curie used *radioactivity* to describe the ability of some elements to give off nuclear radiation.

Nuclear Radiation Is Produced Through Decay (p. 399)

For each type of radiation described in Column A, choose the correct description of what is released in Column B. Write your answer in the space provided.

Column A	Column B
c 10. alpha decay	a. high-energy light
b 11. beta decay	b. positron or electron
a 12. gamma decay	c. helium nucleus

13. Carbon-14 undergoes __beta decay__.
14. Radium-226 undergoes __alpha decay__.

The Penetrating Power of Radiation (p. 401)

15. __Gamma rays__ are the most penetrating of the three forms of nuclear radiation.

16. __Alpha particles__ have the greatest mass of the three forms of nuclear radiation.

Name _____ Date _____ Class _____

Chapter 16, continued

Finding a Date by Decay (p. 403)

24. From birth, the levels of carbon-14 found in an organism increase greatly as the organism grows older and dies.

    True or (False)? (Circle one.)

25. The rate of decay of a radioactive isotope is called __half-life__.

26. Why do you think it is difficult to use carbon-14 to determine the age of extremely old objects (over 50,000 years)?

    Accept any reasonable answer. Sample answer: These objects have less than one one-thousandth of the original carbon-14, which may be too little to measure accurately.

Radioactivity and Your World (p. 405)

27. How are tracers used? (Circle all that apply.)
    a. detecting tumors without using surgery
    b. removing tumors without using surgery
    c. determining fertilizer quality
    d. dating dinosaur bones

28. In Figure 11, how will the engineer detect a leak?

    Using a Geiger counter, the engineer can detect radiation from a tracer added to the oil or gas that leaks out.

Review (p. 405)

Now that you've finished Section 1, review what you learned by answering the Review questions in your ScienceLog.

Section 2: Energy from the Nucleus (p. 406)

1. Why do you think it is important for ordinary citizens to know about the benefits and hazards of using nuclear energy?

   Accept any reasonable answer. Sample answer: Sometimes the public needs to vote on nuclear-power issues.

Name _____ Date _____ Class _____

Chapter 16, continued

17. All of the following are symptoms of radiation sickness *except*
    a. hair loss.
    b. burns.
    c. loss of appetite.
    (d.) rashes.

18. According to the Brain Food on page 401, Marie Curie died of __leukemia__, which was probably caused by exposure to __radiation__.

19. Of the three forms of radiation, __alpha particles__ do the most damage to living cells.

20. Take a moment to read the Environment Connection on page 402. What happens when you breathe radon-222?

    Radon-222 is radioactive, so it will emit alpha particles. If you breathe it, the radiation can damage your lung tissue.

Nuclei Decay to Become Stable (p. 402)

21. The __strong force__ keeps protons and neutrons together in the nucleus of an atom.

22. All nuclei composed of 83 protons or more are radioactive because the strong force
    a. is weakened by electrons.
    b. gets weaker over time.
    (c.) acts only at short distances.
    d. cannot overcome repulsion of the neutrons.

23. What does uranium-238 become after a series of 14 decays takes place?

    It becomes lead-206.

Review (p. 402)

Now that you've finished the first part of Section 1, review what you learned by answering the Review questions in your ScienceLog.

◀◀ CHAPTER 16

Chapter 16, continued

Nuclear Fission (p. 406)

2. In nuclear fission, an atom decays into two **more** stable nuclei. (more or less)

3. Compared to the amount of matter in the nucleus, the small amount of energy that is released when a single uranium nucleus splits is relatively tiny. True or **False** (Circle one.)

4. What is a nuclear chain reaction?

   It is a continuous series of nuclear fission reactions.

5. Nuclear power plants are **less** expensive to operate than conventional power plants. (more or less)

6. The nuclear power plant in Figure 15 uses **uranium-235** nuclei as fuel.

7. Name two hazards associated with the use of nuclear fission to produce power.

   Sample answers: radioactive explosions, radioactive spills, and nuclear waste

Nuclear Fusion (p. 410)

8. Where does fusion occur?
   a. in the Earth's core
   b. on Saturn's surface
   c. in the sun
   d. in modern power plants

9. Look at the Astronomy Connection on page 410. What other substances do stars use as fuel besides hydrogen?

   When stars get older, they use helium, carbon, and silicon as fuel for nuclear fusion.

Chapter 16, continued

10. What is one reason that today's nuclear power plants do not use fusion reactions to produce power?

    Sample answers: Today's technology cannot handle the high temperatures required for nuclear fusion. Also, there is no material on Earth that can hold a hydrogen plasma.

11. What kind of fuel have experimental fusion reactors used?
    a. radon-222
    b. hydrogen-3
    c. uranium-235
    d. helium-4

12. Give three reasons why fusion reactors, in theory, would be better than fission reactors.

    Sample answers: The fuel for fusion is readily available; a fusion reaction produces more energy per gram of fuel; an accident in a fusion reactor would not release much radioactive material; and fusion produces little radioactive waste.

Review (p. 411)
Now that you've finished Section 2, review what you learned by answering the Review questions in your ScienceLog.

CHAPTER 17 DIRECTED READING WORKSHEET

Introduction to Electricity

As you read Chapter 17, which begins on page 420 of your textbook, answer the following questions.

Strange but True! (p. 420)

1. Which of the following is NOT true about electric eels?
   a. An electric eel uses electric discharges to stun or kill its prey.
   b. The body of an adult eel can generate 5,000 to 6,000 V.
   c. The eel's thick skin protects it from electrocuting itself.
   d. Eels swallow their prey whole.

2. The bursts of voltage that an eel gives off when it shocks its prey is greater than the voltage of an electrical outlet. (True)or False? (Circle one.)

What Do You Think? (p. 421)

Answer these questions in your ScienceLog now. Then later, you'll have a chance to revise your answers based on what you've learned.

Investigate (p. 421)

3. What is the purpose of this activity?

   The purpose of this activity is to make electrically charged objects and
   observe their effect on other objects.

Section 1: Electric Charge and Static Electricity (p. 422)

4. When you shuffle your feet on the carpet on a dry day, you get a shock from the metal objects that you touch. What is the cause of this?

   The shock is caused by a buildup of static electricity, which is discharged
   when you touch metal.

Chapter 17, continued

Atoms and Charge (p. 422)

Choose the word in Column B that best matches the description in Column A, and write the corresponding letter in the space provided.

	Column A	Column B
c	5. composed of atoms	**a.** proton
a	6. a positively charged particle of the nucleus	**b.** neutron
d	7. a negatively charged particle found outside the nucleus	**c.** matter
b	8. a particle of the nucleus that has no charge	**d.** electron

9. According to the law of electric charges, like charges attract and opposite charges repel. True or (False) (Circle one.)

10. Why don't electrons fly out of atoms while traveling around the nucleus?

    Protons and electrons have opposite charges, and therefore they attract
    each other.

11. Which of the following does NOT determine the strength of an electric force between charged objects?
    a. the age of the charges
    b. the size of the charges
    c. the distance between the charges
    d. All of the above contribute.

12. The region around a charged particle that can exert a force on another charged particle is called the _electric field_ .

Charge It! (p. 424)

Choose the word in Column B that best matches the definition in Column A, and write the corresponding letter in the space provided.

	Column A	Column B
b	13. "wiping" electrons off of one object onto another	**a.** induction
c	14. transfer of electrons when one object touches another object	**b.** friction
a	15. rearranging electrons in an uncharged object when it is near a charged object	**c.** conduction

Chapter 17, continued

16. When objects are charged, charges cannot be created or destroyed. (**True** or False? (Circle one.)

17. An electroscope can determine which of the following?
    a. whether or not an object is charged
    b. the material that a charged object is made of
    c. the strength of the charge on an object
    d. how many electrons are involved in the charge

Review (p. 426)

Now that you've finished the first part of Section 1, review what you learned by answering the Review questions in your ScienceLog.

Moving Charges (p. 426)

18. Electric cords are often covered in plastic and have metal prongs. This is because metal is a good __conductor__ and plastic is a good __insulator__.

19. Hair dryers should not be used near water. Why? Tap water conducts charges very well, so if you accidentally drop the hair dryer in water, you might receive an electric shock from charges travelling through the water.

Static Electricity (p. 427)

20. What is static electricity?
    a. an electric charge on a stationary object
    b. random electric signals from your dryer
    c. the buildup of electric charges on an object
    d. electricity that moves away from an object

21. As charges move off an object, the object loses its static electricity. This process is called electric __discharge__.

22. How does clothing that has become charged in a dryer lose its charge? The electric charges are transferred to water molecules in the air.

23. An electric discharge can occur quickly or slowly. (**True** or False? (Circle one.)

Chapter 17, continued

24. Standing on a beach or golf course during a thunderstorm can make you like a lightning rod. Why? Beaches and golf courses tend to be flat and have few tall objects to attract lightning. Lightning will be attracted to a person because he or she is the tallest object in the area.

Review (p. 429)

Now that you've finished Section 1, review what you learned by answering the Review questions in your ScienceLog.

Section 2: Electrical Energy (p. 430)

1. Name something that uses electrical energy that would be difficult for you to live without. Explain. Accept any reasonable answer. Sample answer: Refrigerators would be very difficult to live without because keeping food fresh would be more difficult.

Batteries Are Included (p. 430)

Choose the word in Column B that best matches the definition in Column A, and write the corresponding letter in the space provided.

	Column A	Column B
b	2. converts chemical energy into electrical energy	**a.** electrode
e	3. type of cell that contains a solid or pastelike electrolyte	**b.** cell
c	4. mixture of chemicals in a cell	**c.** electrolyte
a	5. where charges enter or exit a cell	**d.** wet
d	6. type of cell that contains a liquid electrolyte	**e.** dry

Bring on the Potential (p. 430)

7. In a battery, electric current exists between the two electrodes because there is a difference in __charge__ between the electrodes.

Chapter 17, continued

8. If the potential difference is increased, the current __increases__ . (increases or decreases)

Other Ways of Producing Electrical Energy (p. 432)

Choose the term in Column B that best matches the description in Column A, and write the corresponding letter in the space provided.

Column A	Column B
__e__ 9. converts kinetic energy into electrical energy	a. silicon
__d__ 10. converts light energy into electrical energy	b. copper
__a__ 11. ejects electrons when struck by light	c. thermocouple
__c__ 12. converts thermal energy into electrical energy	d. photocell
__b__ 13. one type of wire used in some thermocouples	e. generator

Review (p. 432)

Now that you've finished Section 2, review what you learned by answering the Review questions in your ScienceLog.

Section 3: Electric Current (p. 433)

1. Where does most of the electrical energy used in your home come from?
   a. large rechargeable batteries
   (b.) electric power plants
   c. chemical reactions
   d. small generators

Current Revisited (p. 433)

2. In a wire, electrons travel at almost the speed of light.
   True or (False?) (Circle one.)

Label each of the following as a characteristic of alternating current or direct current. Write *AC* for alternating current or *DC* for direct current.

3. __DC__ It's produced by batteries.
4. __AC__ It provides the energy in your home.
5. __AC__ The flow of charges switches directions.
6. __DC__ Charges flow in one direction.
7. __AC__ It's more practical for transferring electrical energy.

Chapter 17, continued

Voltage (p. 434)

8. Another word for potential difference is __voltage__ , which is expressed in __volts__ .

9. As voltage increases in a circuit, the current decreases.
   True or (False?) (Circle one.)

10. The electrical outlets in the United States usually supply a voltage of __120__ V.

11. Look at the Biology Connection on page 435. Why do doctors intentionally create a large voltage across the chest of a heart-attack victim?
    __The large voltage forces the pacemaker cells in the heart to act together, restoring a regular heartbeat.__

Resistance (p. 435)

12. Which of the following associations is false?
    (a.) Increasing resistance increases current.
    b. Decreasing resistance increases current.
    c. Good conductors have low resistance.
    d. Poor conductors have high resistance.

13. An object's resistance depends on which of the following properties of the object? (Circle all that apply.)
    (a.) thickness
    (b.) length
    (c.) temperature
    d. color

14. The light bulb in Figure 18 has a filament made of tungsten. Why is tungsten used in this bulb?
    __Tungsten's high resistance causes electrical energy to be converted to light and thermal energy when charges flow through it.__

15. Thin wires have __more__ resistance than thick wires. Short wires have __less__ resistance than long wires.

16. The resistance of metals generally increases with increasing temperatures because at higher temperatures, the faster-moving atoms slow down the flow of electric charge.
    (True) or False? (Circle one.)

Chapter 17, continued

Ohm's Law: Putting It All Together (p. 437)

17. In the equation for Ohm's law, what do the letters *I*, *V*, and *R* stand for?

    I stands for current, *V* is for voltage, and *R* is for resistance.

18. Who was Georg Ohm?
    a. an electrician   c. an inventor
    b. a teacher   d. an author

19. If you know the current produced in a wire and the voltage applied, you can calculate the resistance of the wire.
    True or False? (Circle one.)

Electric Power (p. 437)

20. Electric power is expressed in
    a. ohms.   c. amperes.
    b. volts.   **d.** watts.

21. Light bulbs may be labeled "100 W" or "40 W." This describes
    a. how long they burn.
    b. how they are disposed of.
    c. how fast the light travels.
    d. how bright they glow.

22. A television uses more power than a hair dryer.
    True or **False**? (Circle one.)

Measuring Electrical Energy (p. 438)

23. In the equation for electrical energy, what do *E*, *P*, and *t* stand for?

    E stands for electrical energy, *P* is for power, and *t* is for time.

24. What do electric meters measure?
    a. power   **c.** current
    b. voltage   **d.** energy

Review (p. 439)

Now that you've finished Section 3, review what you learned by answering the Review questions in your ScienceLog.

Section 4: Electric Circuits (p. 440)

1. An electric circuit always begins and ends in the same place.
   True or False? (Circle one.)

Chapter 17, continued

Parts of a Circuit (p. 440)

2. Which of the following are parts of all electric circuits? (Circle all that apply.)
   a. a load   **c.** wires
   b. an energy source **d.** a switch

3. A ____switch____ opens and closes a circuit.

Types of Circuits (p. 441)

4. A ____series____ circuit has all parts connected in a single loop. A ____parallel____ circuit has different loads on separate branches.

5. A string of holiday lights wired together in series has a burned-out bulb. Why do all of the lights go out?

   The burned-out bulb is a break in the circuit, so the circuit is no longer a closed path through which electric charges may flow.

6. If a break occurs in one of the loops of a parallel circuit, current will not run through the other loops. True or **False**? (Circle one.)

Household Circuits (p. 444)

7. Which of the following may cause a circuit failure? (Circle all that apply.)
   a. water   **c.** too many loads
   b. broken wires   **d.** excess insulation

8. As more loads are added to a parallel circuit, the entire circuit draws more current. **True** or False? (Circle one.)

9. How does a fuse disrupt the flow of charges when the current is too high?
   a. A metal strip warms up and bends away from the circuit wires.
   b. A metal strip warms up and melts, leaving a gap.
   c. A metal strip changes from a conductor to an insulator.
   d. None of the above

10. Circuit breakers are inconvenient because breakers must be replaced when they are tripped. True or **False**? (Circle one.)

Review (p. 445)

Now that you've finished Section 4, review what you learned by answering the Review questions in your ScienceLog.

CHAPTER 18 DIRECTED READING WORKSHEET

Electromagnetism

As you read Chapter 18, which begins on page 452 of your textbook, answer the following questions.

Would You Believe . . . ? (p. 452)

1. How do scientists use magnets to make a frog float in midair?

   To make a frog float in midair, scientists expose the frog to a strong magnet.

   The magnetic force between the frog's atoms and the magnet lifts the frog

   and makes it float.

What Do You Think? (p. 453)

Answer these questions in your ScienceLog now. Then later, you'll have a chance to revise your answers based on what you've learned.

Investigate! (p. 453)

2. When you drag a bar magnet down a nail, what must you be sure to do?

   I should drag the magnet in only one direction.

Section 1: Magnets and Magnetism (p. 454)

3. Which of the following statements is NOT true about magnets?
   a. Magnets can stick to some objects without touching them.
   b. Magnets can stick to all types of metals.
   c. Magnets can stick to other magnets.
   d. Magnets are sometimes strong enough to hold up paper on a refrigerator door.

Properties of Magnets (p. 454)

4. _Magnetite_, a mineral that attracts objects containing iron, was found by the Greeks in a part of Turkey called Magnesia over _2,000_ years ago.

5. The poles of a magnet are the parts of a magnet where the magnetic effects are strongest. (**True** or False? (Circle one.))

Chapter 18, continued

6. If you move the south poles of two magnets together, what will happen? Explain.

   The magnetic force will push the poles apart because like poles repel.

7. Take a look at Figure 4 on page 456. What do iron filings do when you sprinkle them around a magnet?

   The filings form curved lines similar to magnetic field lines.

What Makes Materials Magnetic? (p. 456)

8. In some materials, the magnetic fields of the atoms are so _strong_ that they group together in small, magnetlike regions called domains. (weak or **strong**)

9. According to the Brain Food at the bottom left of page 456, how do ranchers protect their cows with magnets?
   a. The cows wear magnets around their necks.
   b. The magnets are ground into a powder and applied to the cows' hides.
   c. The magnets are attached to a barbed wire fence.
   d. The cows swallow magnets that remain in their stomachs.

10. Why would a magnet lose its magnetic properties if it were dropped?

    The impact of being dropped can jostle the domains out of alignment.

11. When you hold a magnet close to an unmagnetized object, the _domains_ in the object align to create a _temporary_ magnet.
    (atoms or domains; temporary or permanent)

12. When you cut a magnet in half, you end up with one north-pole piece and one south-pole piece. True or (**False**) (Circle one.)

Review (p. 458)

Now that you've finished the first part of Section 1, review what you learned by answering the Review questions in your ScienceLog.

Chapter 18, continued

Types of Magnets (p. 458)

Choose the type of magnet in Column B that best matches the description in Column A, and write your answer in the space provided.

Column A	Column B
c 13. easy to create but loses its magnetism easily	a. ferromagnet
d 14. difficult to magnetize but retains its magnetic properties well	b. electromagnet
b 15. produced by an electric current	c. temporary magnet
a 16. made from iron, nickel, or cobalt	d. permanent magnet

Earth as a Magnet (p. 459)

Mark each of the following statements *True* or *False*.

17. __True__ Earth's magnetic poles are not the same as Earth's geographic poles.
18. __False__ The magnetic field lines around Earth are very different from those around a bar magnet.
19. __True__ The magnetic south pole is at the North Pole.
20. __True__ The auroras occur where Earth's magnetic field bends inward at the magnetic poles.

Review (p. 461)

Now that you've finished Section 1, review what you learned by answering the Review questions in your ScienceLog.

Section 2: Magnetism from Electricity (p. 462)

1. Why does a maglev train float?
   The train floats because of the magnetic forces between the track and the train.

The Discovery of Electromagnetism (p. 462)

2. Oersted discovered that a wire carrying a(n) __electric current__ can move a compass needle.

3. The direction of a magnetic field depends on the direction of the __current__ that produced it.

Chapter 18, continued

Using Electromagnetism (p. 463)

4. How can you strengthen a solenoid's magnetic field?
   You can strengthen the magnetic field by adding more loops of wire to the coil and by increasing the current in the wire.

5. What makes up an electromagnet?
   An electromagnet is a magnet made up of a solenoid wrapped around an iron core.

6. Some electromagnets are strong enough to lift a car.
   (**True**) or False? (Circle one.)

Magnetic Force and Electric Current (p. 465)

7. A magnet causes a current-carrying wire to move by exerting a __force__ on the wire.

Applications of Electromagnetism (p. 465)

8. How does a solenoid cause your doorbell to ring?
   Current traveling through the solenoid creates a magnetic field. This field pulls the iron rod through the solenoid, and the rod strikes the bell.

9. An electric motor changes electrical energy into __kinetic__ energy. (kinetic or static)

Mark each of the following statements *True* or *False*.

10. __False__ A commutator is a loop or coil of wire in an electric motor that can rotate.
11. __False__ Devices used by electricians may contain a galvanometer, which measures voltage.

Review (p. 467)

Now that you've finished Section 2, review what you learned by answering the Review questions in your ScienceLog.

ANSWER KEY

Section 3: Electricity from Magnetism (p. 468)

1. What process do power companies use to supply your home with electrical energy?

   To supply my home with electrical energy, power companies use the _____ process of producing an electric current with a magnetic field.

Electric Current from a Magnetic Field (p. 468)

2. In his setup using an iron ring, Faraday discovered that a very strong electromagnet could induce an electric current in the second wire. True or (False)? (Circle one.)

3. When did the pointer on Faraday's galvanometer move?
   a. the instant the galvanometer was connected to or disconnected from the battery
   b. as long as the galvanometer was connected to the battery
   c. as long as the galvanometer was disconnected from the battery

4. What process did Faraday discover during his experiment with the iron ring? Explain.

   Faraday discovered the process of electromagnetic induction. He discovered that an electric current is induced in a wire as long as the magnetic field around the wire is changing.

5. Look at Figure 21 on page 469. Which of the following actions would induce a larger current when a magnet moves through a coil of wire?
   a. moving the magnet more slowly
   b. adding more loops of wire
   c. using thicker wire
   d. reversing the direction the magnet moves

6. An electric current is produced in a wire when the wire moves _____ across _____ magnetic field lines.
   (across or along)

Chapter 18, continued

Applications of Electromagnetic Induction (p. 470)

7. A generator converts kinetic energy into _____ electrical _____ energy. (electrical or mechanical)

8. Generators in power plants produce direct current. True or (False)? (Circle one.)

9. Using the text on page 471, place the following steps of creating electrical energy from nuclear energy in order of occurrence. Write the appropriate number in the space provided.

   5 An electric current is induced, producing electrical energy.
   3 Steam turns a turbine.
   1 A nuclear reaction occurs and creates thermal energy.
   4 The generator's magnet is turned.
   2 Water is boiled to produce steam.

10. What's the difference between a step-up transformer and a step-down transformer?

    A step-up transformer has more loops in the secondary coil and increases voltage. A step-down transformer has fewer loops in the secondary coil and decreases voltage.

11. Electric current is distributed to your house at the same voltage that it is produced by power plants. True or (False)? (Circle one.)

Review (p. 473)

Now that you've finished Section 3, review what you learned by answering the Review questions in your ScienceLog.

CHAPTER 19 DIRECTED READING WORKSHEET

Electronic Technology

As you read Chapter 19, which begins on page 480 of your textbook, answer the following questions.

Would You Believe . . . ? (p. 480)

1. What are three differences between the first televisions and modern televisions?

   Student answers should include three of the following: The first televisions were bulkier, more expensive, and had smaller screens and fuzzier pictures than modern televisions. In addition, the first televisions only produced black and white images.

2. What are two advances you'd like to see in the TVs of the future?

   Accept any reasonable answer. Sample answer: I'd like to see TVs that are voice-controlled and solar-powered.

What Do You Think? (p. 481)

Answer these questions in your ScienceLog now. Then later, you'll have a chance to revise your answers based on what you've learned.

Investigate! (p. 481)

3. The purpose of this activity is to make a model of a __telephone__.

4. In this activity, do you think the length of the string matters? Explain.

   Sample answer: Yes; I think the strength of the sound would decrease as the length of the string increased.

Section 1: Electronic Components (p. 482)

5. Electronic devices change electrical energy into thermal energy. True or (False) (Circle one.)

Inside an Electronic Device (p. 482)

6. Figure 1 shows the _____ inside a TV remote control.
   a. motor
   b. buttons
   (c.) circuit board
   d. TV antenna

7. An LED, or __light-emitting diode__, is one of the electronic components within a TV remote control.

8. The LED in a remote control
   a. gives off radio waves.
   b. is only for decoration.
   (c.) sends information to the TV.
   d. receives information from the TV.

9. Electronic components control electric current. (True) or False? (Circle one.)

Semiconductors (p. 483)

10. If a substance conducts __electrical__ energy better than a(n) __insulator__ but not as well as a conductor, then the substance is called a semiconductor.

11. In pure silicon,
    a. no valence electrons are shared between atoms.
    (b.) there are no free electrons to create electric current.
    c. there are some atoms of gallium.
    d. there are ten valence electrons.

12. Use Figure 3 to explain why arsenic-doped silicon conducts electric current better than pure silicon.

    Arsenic-doped silicon conducts electric current better than pure silicon because arsenic has five valence electrons. When an atom of arsenic replaces an atom of silicon, there is an "extra" electron. The extra electrons conduct electric current.

13. A(n) __p-type__ semiconductor is a semiconductor with "holes." (n-type or p-type)

Name _____ Date _____ Class _____

Chapter 19, continued

Diodes and Transistors (p. 484–486)

After reading pp. 484–486, decide whether each of the following statements describes a diode or a transistor. In the space provided, write D if the statement describes a diode and T if it describes a transistor.

14. __T__ has three "legs" that conduct electric current
15. __T__ can be used as a switch
16. __D__ a two-layer semiconductor sandwich
17. __T__ can be used to amplify electric current
18. __D__ can be used to convert AC to DC
19. __D__ allows current in only one direction
20. __T__ a three-layer semiconductor sandwich

21. Many transistors and diodes can be contained in one integrated circuit. (True) or False? (Circle one.)

22. How have integrated circuits affected the size of electronic devices? Explain.

An integrated circuit can contain many complete circuits in a very small space, so integrated circuits have allowed electronic devices to become much smaller.

23. Which of the following statements is NOT true about vacuum tubes?
    a. They were invented before transistors.
    (b.) They could not amplify electric current.
    c. They can be used to change AC to DC.
    d. They were used in early radios.

Name _____ Date _____ Class _____

Chapter 19, continued

24. What are two reasons transistors have replaced vacuum tubes in many modern electronic devices?

Answers will vary but should include two of the following: transistors are smaller than vacuum tubes, transistors give off less thermal energy than vacuum tubes, and transistors last longer than vacuum tubes.

Review (p. 487)

Now that you've finished Section 1, review what you learned by answering the Review questions in your ScienceLog.

Section 2: Communication Technology (p. 488)

1. The first telecommunication device was the
    a. telephone.   c. fax.
    b. television.   (d.) telegraph.

2. Using the table in Figure 10, on page 488, write your name in International Morse Code in the space provided below.

Answers will vary. The answer should be the student's name written in dots and dashes.

Communicating with Signals (p. 488)

3. A signal is anything, such as a sound or a series of numbers, that represents _____ information _____.

4. A signal may travel better inside a _____ carrier _____, which is another form of energy.

Analog Signals (p. 489)

5. Read the Geology Connection. Since a seismogram is an analog signal, it shares none of the properties of the waves produced by an earthquake. True or (False)? (Circle one.)

Chapter 19, continued

6. Your telephone's analog signal is a wave of **electric current**. (sound or electric current)

7. Use Figure 11 to place the following events in the correct order to explain how a telephone works. Write the appropriate number in the space provided.

   4 Electric current is converted to a sound wave.

   2 Vibrations are converted to electric current.

   1 Sound waves enter the transmitter and vibrate a metal disk.

   3 An analog signal travels through the phone wires.

8. In vinyl records, the number and depth of the contours in the grooved walls represent the **frequency** and **loudness** of the sounds.

9. The stylus riding in the grooves of a record causes an electromagnet to vibrate. (True)or False? (Circle one.)

Digital Signals (p. 490)

Choose the word or phrase in Column B that best matches the description in Column A, and write your answer in the space provided.

Column A	Column B
d 10. represented by a pulse	a. a digital signal
c 11. represented by a missing pulse	b. binary
a 12. a series of electrical pulses representing the digits of binary numbers	c. the number 0
b 13. means "two"	d. the number 1

14. Why are the pits and lands on a CD important?

    Pits and lands store information so it can be converted by a CD player into sound.

15. The higher the sampling rate, the better the digital signal reproduces the original sound wave. (True)or False? (Circle one.)

Chapter 19, continued

16. Why do CDs last so long?

    CDs last so long because they are read by a laser beam, which does not wear down the CD.

Radio and Television (p. 492)

17. Waves that travel as changing electric and magnetic fields are called **electromagnetic** waves.

18. Look at Figure 16. A modulator

    a. changes sound waves into electric current.
    b. strengthens the analog signal.
    c. combines the analog signal with radio waves.
    d. removes the radio waves from the analog signal.

Mark each of the following statements *True* or *False*.

19. **True** Electrons are accelerated toward your TV screen by special guns.

20. **False** In television, only audio signals are carried by radio waves.

21. **True** Fluorescent materials glow red, blue, and green when beams of electrons strike the TV screen.

Review (p. 493)

Now that you've finished Section 2, review what you learned by answering the Review questions in your ScienceLog.

Section 3: Computers (p. 494)

What Is a Computer? (p. 494)

1. The alarm clock pictured in Figure 18 is a computer. (True)or False? (Circle one.)

2. Computers can think. True or (False) (Circle one.)

3. Which of the following are the basic functions performed by a computer? (Circle all that apply.)

   a. processing   **d.** output
   b. input   **e.** programming
   c. deleting   **f.** storage

Name _____ Date _____ Class _____

Chapter 19, continued

Historic Developments (p. 495)

4. Which of the following statements is NOT true of the ENIAC?
   a. It was built with thousands of vacuum tubes.
   b. It was very inexpensive to build.
   c. It was the first general-purpose computer.
   d. It was developed by the United States Army.

5. Integrated circuits have made today's computers small and powerful. (**True** or False? (Circle one.)

Computer Hardware (p. 496)

Read pp. 496–497. Then choose the part of the computer in Column B that best matches the example or definition in Column A, and write your answer in the space provided. Some parts may be used more than once.

Column A	Column B
d 6. a monitor	a. input device
a 7. where information is given to a computer	b. central processing unit
a 8. a keyboard	c. memory
b 9. where the computer does calculations	d. output device
c 10. RAM	
c 11. where a computer stores data	
d 12. where a computer shows its results	
a 13. your voice	

14. What is the difference between RAM and ROM?
ROM is permanent memory that cannot be added to or changed. RAM is temporary memory that stores information only while the information is being used.

15. What do computers use to "talk" with each other?
   a. modems
   b. speakers
   c. CD-ROMs
   d. scanners

Name _____ Date _____ Class _____

Chapter 19, continued

Computer Software (p. 498)

16. Software is a set of _____ instructions _____ that makes it possible for a computer to perform a task.

Mark each of the following statements *True* or *False*.

17. ___True___ Operating system software interprets commands from the input device.

18. ___True___ A word processor is an example of a utility.

19. ___False___ Application software supervises the interactions of the software and the hardware.

The Internet—A Global Network (p. 499)

20. Internet connections at home and at school are exactly the same. True or (**False**)? (Circle one.)

21. Imagine that you are at home communicating over the Internet with someone in Rome, Italy. Describe how you are connected.
My computer connects to an Internet Service Provider, or ISP, over a phone line. The ISP is connected by a satellite to an ISP in Europe, which in turn is connected to the computer used by the person in Rome.

Review (p. 499)

Now that you've finished Section 3, review what you learned by answering the Review questions in your ScienceLog.

Name _____ Date _____ Class _____

CHAPTER 20 DIRECTED READING WORKSHEET

The Energy of Waves

As you read Chapter 20, which begins on page 508 of your textbook, answer the following questions.

This Really Happened! (p. 508)

1. How did a land-based earthquake cause the worst marine disaster in the history of the town of Kodiak?

   Sample answer: The earthquake created a series of waves called tsunamis in the Gulf of Alaska. When the waves reached the shore, they pounded everything in their path.

What Do You Think? (p. 509)

Answer these questions in your ScienceLog now. Then later, you'll have a chance to revise your answers based on what you've learned.

Investigate! (p. 509)

2. What is the purpose of this activity?

   The purpose of this activity is to observe and predict the action of waves.

Section 1: The Nature of Waves (p. 510)

3. Which of the following waves might your family have experienced after a day at the beach? (Circle all that apply.)
   a. water waves
   b. microwaves
   c. light waves
   d. sound waves

Waves Carry Energy (p. 510)

4. A wave can carry energy away from its source. (True) or False? (Circle one.)

DIRECTED READING WORKSHEETS **145**

Name _____ Date _____ Class _____

Chapter 20, continued

5. Take a moment to examine Figure 1. Floating birds and boats bob up and down and travel in the same direction as waves. True or (False)? (Circle one.)

6. Air doesn't travel with sound waves. Give an example of what would happen if it did.

   Sample answer: If air traveled with sound, I would feel a rush of air every time the phone rang.

7. Waves can transfer _____ energy _____ through the vibration of _____ particles _____ in solid, liquid, or gaseous media.

8. If you put an alarm clock inside a jar and remove all the air from the jar, you wouldn't hear the alarm ringing. Why?

   Sample answer: I wouldn't hear the clock ringing because sound energy travels through the vibration of air particles. Without air particles to vibrate, no sound is possible.

9. Electromagnetic waves are waves that require a medium to transfer energy. True or (False) (Circle one.)

10. Take a moment to look at the Astronomy Connection in the left column of page 512. In what way is looking at the light from a star like viewing the past?

    The light from a star takes many years to arrive on Earth. The light that I see shows me what the star looked like years ago, when the light was just emitted from the star.

11. Which of the following lists of waves contains ONLY mechanical waves?
    a. radio waves, X rays, sound waves
    b. seismic waves, water waves, microwaves
    c. microwaves, water waves, X rays
    (d.) sound waves, seismic waves, water waves

146 HOLT SCIENCE AND TECHNOLOGY

ANSWER KEY

Name _____ Date _____ Class _____

Chapter 20, continued

Types of Waves (p. 513)

After reading pages 513–515 in your text, match each type of wave in Column A to the correct statements in Column B, and write the corresponding letter in the appropriate space. Wave types can be used more than once.

Column A	Column B
c 12. Particles move forward at the crest and backward at the trough.	a. longitudinal wave
b 13. Particles vibrate up and down.	b. transverse wave
a 14. Electromagnetic waves are this type of wave.	c. surface wave
a 15. A rarefaction is a section of this type of wave that is less crowded than normal.	
a 16. Particles vibrate back and forth along the path the wave travels.	
a 17. A sound wave is this type of wave.	
c 18. Two types of waves can combine to form this type of wave.	
b 19. Particles travel perpendicular to the direction the wave travels.	
c 20. This wave occurs at or near the boundary of two media.	

Review (p. 515)

Now that you've finished Section 1, review what you learned by answering the Review questions in your ScienceLog.

Section 2: Properties of Waves (p. 516)

1. Waves made by the breeze were very different than waves created by the speedboat. Describe the difference.

Sample answer: The waves made by the breeze were short and close together. The waves created by the speedboat were tall and widely-spaced.

Name _____ Date _____ Class _____

Chapter 20, continued

Amplitude (p. 516)

2. Which of the following is NOT true of amplitude?
   a. The smaller the amplitude, the more energy that is carried by a wave.
   b. It is the maximum distance a wave vibrates from its resting position.
   c. The larger the amplitude of a water wave, the taller the wave.
   d. All of the above are true.

(a is circled)

Wavelength (p. 517)

3. Wavelength can be measured between corresponding points on two adjacent waves. (True or False? (Circle one.)

(True is circled)

Frequency (p. 518)

4. The frequency of a wave is the _____ number _____ of waves produced in a given amount of time. To measure frequency of a transverse wave, I can count the number of _____ crests _____ or _____ troughs _____ that pass a point in a certain amount of time. I can also count the number of _____ compressions _____ or _____ rarefactions _____ for a longitudinal wave.

Mark each of the following statements True or False.

5. __False__ The frequency of a wave is not related to its wavelength.
6. __True__ When wavelength decreases, frequency increases.
7. __True__ A wave with a low frequency carries less energy than a wave with a high frequency.
8. __True__ Frequency is expressed in hertz (Hz).

Wave Speed (p. 519)

9. The speed of a wave is the _____ distance _____ traveled by a wave in a given amount of _____ time _____. Wave speed is equal to _____ wavelength _____ times _____ frequency _____.

Chapter 20, continued

10. Complete the MathBreak in your ScienceLog. Then answer the following questions:
   a. What can you calculate if you know the frequency and speed of a wave?

   I can calculate its wavelength.

   b. If you know the speed of a wave, and you want to find its wavelength, what other piece of information would be helpful for you to know?

   It would be helpful for me to know the frequency of the wave.

Review (p. 519)

Now that you've finished Section 2, review what you learned by answering the Review questions in your ScienceLog.

Section 3: Wave Interactions (p. 520)

1. The moon doesn't produce light like the stars do. So why does the moon shine?

   The moon shines because light from the sun reflects off its surface.

Reflection (p. 520)

2. Reflection is when waves hit a barrier and some of them pass through it. True or (False)? (Circle one.)

3. Water waves cannot be reflected. True or (False)? (Circle one.)

4. Sound waves that reflect off canyon walls or classroom walls are called _____ echoes _____.

Chapter 20, continued

Refraction (p. 521)

5. Why does the pencil in a half-filled glass of water look like it's broken?

   The light waves entering the water are refracted, and this causes us to see the pencil differently in the water.

6. A wave bends when it enters a new medium because the part of the wave that enters first is traveling at a different speed than the rest of the wave. (True) or False? (Circle one.)

Diffraction (p. 521)

7. Which of the following are true of diffraction? (Circle all that apply.)
   a. Waves curve or bend when they reach the edge of an object.
   b. Waves bend through an opening.
   c. Sound travels around buildings.
   d. The size of the barrier does not affect the amount of diffraction.

8. Diffraction of a wave depends on what two things?

   The diffraction of a wave depends on the wavelength and on the size of the barrier or opening.

9. Look at Figure 18 on page 522. If the opening is larger than the wavelength, _____ a little _____ diffraction occurs. (a little or a lot of)

Interference (p. 522)

10. Which of the following is TRUE about overlapping waves?
    a. They share the same space.
    b. They cannot be in the same place at the same time.
    c. They pass around each other.

11. Interference is the result of two or more waves combining to form three waves. True or (False)? (Circle one.)

Chapter 20, continued

Match each type of interference in Column B to the correct statements in Column A, and write the corresponding letter in the appropriate space. Interference types can be used more than once.

Column A	Column B
_a_ 12. The crests of two waves overlap.	a. constructive interference
_b_ 13. The resulting wave has a smaller amplitude than the original waves had.	b. destructive interference
_b_ 14. Waves of the same amplitude cancel each other out.	
_a_ 15. The result is a wave with deeper troughs and higher crests than the original waves.	
_b_ 16. Crests of one wave overlap with the troughs of another wave.	

17. A standing wave occurs from interference between the original wave and the reflected wave. (True) or False? (Circle one.)

18. In a standing wave, ___constructive___ interference causes portions of the wave to have a large amplitude, while ___destructive___ interference causes other ___total___ portions of the wave to be at rest.

19. Standing waves only *look* like they're standing still. What's really going on?
    a. Waves are vibrating faster than light.
    b. Waves are no longer vibrating.
    (c.) Waves are traveling in both directions.

20. Place the following steps in the correct sequence to explain how the marimba player in Figure 23 uses resonance. Write the appropriate number in the space provided.
    _6_ The amplitude of the vibrations is increased.
    _2_ The bar vibrates.
    _1_ The marimba player strikes a bar.
    _7_ A loud note is produced.
    _3_ The air in the column underneath the bar absorbs energy from the vibrating bar and begins to vibrate.
    _4_ The frequency of the air column matches the frequency of the bar.
    _5_ The air column resonates with the bar.

Chapter 20, continued

21. During resonance, a vibrating object will cause a second object to vibrate when it reaches the second object's resonant frequency. (True) or False? (Circle one.)

22. How did the Tacoma Narrows Bridge earn the nickname Galloping Gertie?
    a. The bridge was used most often by people traveling on horseback.
    (b.) The bridge experienced wavelike motions during strong winds.
    c. The bridge was built by Gertrude Stein in July 1940.
    d. Soldiers marched across it in such a way that it collapsed in 1831.

23. How did resonance contribute to the destruction of Galloping Gertie?
    Sample answer: Wind that blew across the bridge caused the bridge to vibrate at a frequency close to the resonant frequency of the bridge. The bridge absorbed large amounts of energy from the wind, which caused the bridge to vibrate with a large amplitude. When a cable slipped, the bridge started to twist, increasing the amplitude. Eventually, the amplitude became large enough to cause the bridge to collapse.

Review (p. 525)

Now that you've finished Section 3, review what you learned by answering the Review questions in your ScienceLog.

CHAPTER 21

DIRECTED READING WORKSHEET

The Nature of Sound

As you read Chapter 21, which begins on page 532 of your textbook, answer the following questions.

Would You Believe . . . ? (p. 532)

1. What did Marco Polo see and hear in Asia that amazed him so much?

   Marco Polo saw and heard the booming sand dunes of the Asian desert.

2. Which of the following are true about the noises made by the sands? (Circle all that apply.)
   a. They can be heard more than 50 km away.
   b. People have compared the sound to foghorns, moaning, and cannon fire.
   c. They don't occur in the United States.
   d. They tend to be found in large deserts.

What Do You Think? (p. 533)

Answer these questions in your ScienceLog now. Then later, you'll have a chance to revise your answers based on what you've learned.

Investigate! (p. 533)

3. What is the purpose of this activity?

   The purpose of this activity is to explore the nature of sound using a home-made guitar.

Section 1: What Is Sound? (p. 534)

4. What are two sounds you hear indoors?

   Accept all reasonable answers. Sample answers: people talking, radio blaring, or dishes clattering

Sound Is Produced by Vibrations (p. 534)

5. The complete back-and-forth motion of an object is called a vibration.

Chapter 21, continued

Look at Figure 1 to answer items 6 and 7.

6. In a compression, the molecules in the air are **more** closely packed than in the surrounding air. (more or less)

7. In a rarefaction, the molecules in the air are **less** closely packed than in the surrounding air. (more or less)

8. Sound waves are an example of **longitudinal** waves. (transverse or longitudinal)

9. Take a moment to read the Biology Connection on page 535. What causes the sounds that you make when you speak? (Circle all that apply.)
   a. air making your vocal cords vibrate
   b. air rushing down your windpipe
   c. air vibrating in your chest
   d. air forced through your windpipe

10. Air does not travel with sound waves. But what would happen at the school dance if air did travel with sound?

    Wind gusts from the speakers would knock people over.

Creating Sounds Vs. Detecting Sound (p. 535)

11. What happens to the surrounding air when a tree falls and hits the ground?

    Vibrations in the tree and ground create compressions and rarefactions in the air.

Sound Waves Require a Medium (p. 536)

12. In the example above, the medium through which sound travels is the ground. True or **False**? (Circle one.)

13. If a tree fell in a vacuum, why wouldn't you hear a sound?

    Sample answer: In a vacuum there are no air particles to vibrate, and it is the vibration of air particles that produces sound.

Chapter 21, continued

14. Take a moment to read the Astronomy Connection on page 536. Name one kind of wave that can travel in a vacuum.

Radio waves can travel without a medium.

How You Detect Sound (p. 536)

15. After your ears convert sound waves into electrical signals, where are the signals sent for interpretation?

a. pinna   c. brain
b. spinal cord d. oval window

16. Place each of the following terms in the correct column: *hammer, stirrup, oval window, cochlea, hair cells, pinna, eardrum, ear canal, anvil*. (If the term refers to an entrance to a part of the ear, write the term between the appropriate columns.)

Outer ear		Middle ear		Inner ear
pinna		hammer	oval window	cochlea
ear canal	eardrum	anvil		hair cells
		stirrup		

Choose the term in Column B that best matches the description in Column A, and write your answer in the space provided.

Column A	Column B
_b_ **17.** the outermost portion of the ear	a. cochlea
_e_ **18.** bends to stimulate nerves	b. pinna
_a_ **19.** portion of the ear that contains liquid	c. hammer
_d_ **20.** the bone that vibrates the oval window	d. stirrup
_c_ **21.** the eardrum makes this bone vibrate	e. hair cell

22. What ways can you protect yourself from tinnitus? (Circle all that apply.)

a. drinking a glass of milk each day
b. wearing ear protection while operating heavy machinery
c. getting lots of sleep
d. turning your radio down

Review (p. 538)

Now that you've finished Section 1, review what you learned by answering the Review questions in your ScienceLog.

Chapter 21, continued

Section 2: Properties of Sound (p. 539)

Name an example of each of the following from your everyday life. Accept all reasonable answers. Samples are given.

1. a soft sound: _a person whispering_
2. a loud sound: _a jackhammer being used_
3. a high-pitched sound: _a bird chirping_
4. a low-pitched sound: _a person playing the tuba_

The Speed of Sound Depends on the Medium (p. 539)

5. How quickly a sound reaches your ears depends on how loud it is. True or **False**? (Circle one.)

6. How quickly a sound reaches your ears depends on the medium through which the sound is traveling. **True** or False? (Circle one.)

7. What did Chuck Yeager accomplish in 1947?

He became the first person to fly faster than the speed of sound.

8. In general, as the medium cools, the speed of sound _decreases_ . (increases or decreases)

9. When particles slow down

a. they transmit energy more quickly.
b. they transmit energy more slowly.
c. they gain kinetic energy.
d. None of the above

Pitch Depends on Frequency (p. 540)

10. In a guitar, pitch is NOT related to

a. the frequency of the sound wave.
b. the thickness of the guitar string.
c. the number of sound waves produced in a given time.
d. how far away the guitar is from your ear.

Name _____ Date _____ Class _____

Chapter 21, continued

▶▶ CHAPTER 21

11. Of the animals shown in the graph on page 541:
    a. Which animal can hear sounds with the highest pitch?
       a bat
    b. Which animal can hear sounds with the lowest pitch?
       a dog

12. What frequencies of sound are infrasonic?
    Sounds with frequencies lower than 20 Hz are infrasonic.

13. Name two applications of ultrasonic waves.
    Sample answers: cleaning jewelry; removing ice from airplane wings, car windshields, and freezers; and breaking kidney stones into smaller pieces

14. Look at the Biology Connection on page 541. What is a kidney stone?
    a. a bone inside your kidney
    b. a kidney-shaped rock used in ultrasound
    c. an ancient tool used to harvest kidney beans
    d. a calcium-salt deposit in kidneys

15. The Doppler effect does NOT affect how sound is perceived by
    a. a pedestrian when a honking driver speeds by.
    b. a driver who honks while speeding by a pedestrian.
    c. a person sitting in a parked car when a driver honks as he speeds by.
    d. a driver speeding past a person who honks the horn of his parked car.

DIRECTED READING WORKSHEETS 157

Name _____ Date _____ Class _____

Chapter 21, continued

Loudness Is Related to Amplitude (p. 542)

Choose the term in Column B that best matches the definition in Column A. Write your answer in the space provided.

Column A	Column B
_a_ **16.** the unit used to express how loud or soft a sound is perceived	**a.** decibel
_b_ **17.** how loud or soft a sound is perceived	**b.** loudness
_c_ **18.** the maximum distance the particles in a wave vibrate from their rest positions	**c.** amplitude

Use the three diagrams of sounds graphed on the oscilloscopes below to answer questions 19–23.

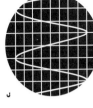

a.   b.   c.

19. Which has the lowest frequency? _b_
20. Which has the smallest amplitude? _b_
21. Which has the highest frequency? _a_
22. Which has the largest amplitude? _c_
23. In oscilloscope (b) above, does the left edge of the graph represent a compression, a rarefaction, or neither?
    It represents a compression.

Review (p. 544)
Now that you've finished Section 2, review what you learned by answering the Review questions in your ScienceLog.

158 HOLT SCIENCE AND TECHNOLOGY

Name _____ Date _____ Class _____

Chapter 21, continued

Section 3: Interactions of Sound Waves (p. 545)

1. List two reasons why sounds are important to beluga whales.
   Sample answer: They use sounds to communicate and to find food.
   _____

Reflection of Sound Waves (p. 545)

2. A hard, rigid surface is a better reflector of sound than a soft surface is. (**True** or False? (Circle one.)

3. A shout in a gymnasium will usually produce more echoes than a shout in an auditorium. (**True** or False? (Circle one.)

4. In Figure 13, how does the Doppler effect help the bat find food?
   Sample answer: Because of the Doppler effect, the frequency of the echo tells the bat if the insect is flying toward it.
   _____

5. When do you think it is best for ships to use sonar technology instead of just relying on eyesight to locate icebergs?
   Accept any reasonable answer. Sample answer: It is best to use sonar technology during a storm or at night, when visibility is poor.
   _____

6. Which of the following ways can ultrasonic waves be used in medicine? (Circle all that apply.)
   a. performing surgery without making an incision
   b. monitoring the development of an unborn baby
   c. detecting skin cancer
   d. examining internal organs

Review (p. 547)

Now that you've finished the first part of Section 3, review what you learned by answering the Review questions in your ScienceLog.

Name _____ Date _____ Class _____

Chapter 21, continued

Interference of Sound Waves (p. 548)

7. In ___destructive___ interference, two sound waves can overlap to produce a softer sound, while in ___constructive___ interference, two sound waves combine to produce a louder sound.

8. A sonic boom is created when an airplane travels faster than the speed of ___sound___ . (light or sound)

9. Using Figure 18, explain what happens when a jet flies at supersonic speeds.
   When a jet flies faster than the speed of sound, the sound waves spread out behind it in a cone shape. The sound waves combine on the edge of the cone by constructive interference. This creates a shock wave, which you hear as a sonic boom.

10. In a standing wave, portions of the wave are ___at rest___ while other portions have a larger amplitude due to interference.

11. Suppose you are using a tuning fork to cause a guitar string to vibrate without touching it. You get the best results when the resonant frequency of the tuning fork matches the
    a. fundamental frequency of the string.
    b. first overtone frequency of the string.
    c. second overtone frequency of the string.
    d. third overtone frequency of the string.

Diffraction of Sound Waves (p. 551)

12. Sound waves with a high frequency have a ___short___ wavelength and ___are not___ easily diffracted.
    (long or short, are or are not)

13. When a radio is playing in the next room, which sound waves can be heard the best?
    a. sound waves with a high frequency
    b. sound waves with a low frequency
    c. sound waves with a high pitch
    d. sound waves with a small amplitude

ANSWER KEY

Chapter 21, continued

Section 4: Sound Quality (p. 552)

Review (p. 551)
Now that you've finished Section 3, review what you learned by answering the Review questions in your ScienceLog.

1. What is the difference between music and noise?
   a. loudness   c. amplitude
   b. pitch   **d. sound quality**

What Is Sound Quality? (p. 552)

2. Why do the same notes sound different on different instruments?

   Each instrument has a unique sound quality, or blending together of several pitches due to interference.

Sound Quality of Instruments (p. 553)

3. Place each of the following instruments in the correct column: *drum, guitar, trumpet, cello, bells, tuba, clarinet, banjo, cymbals, violin, saxophone.*

String instruments	Wind instruments	Percussion instruments
guitar	trumpet	drum
cello	tuba	bells
banjo	clarinet	cymbals
violin	saxophone	

Fill in each blank in items 4–6 using either *lower pitch* or *higher pitch*.

4. In a string instrument, a longer string has a _lower pitch_ and a thinner string has a _higher pitch_.

5. In a wind instrument, shortening the air column produces a _higher pitch_.

6. Among percussion instruments, smaller instruments produce a _higher pitch_.

Chapter 21, continued

Music or Noise? (p. 555)

7. Which of the following would produce a sound with a random mix of frequencies? (Circle all that apply.)
   a. slamming door
   b. French horn
   c. truck engine
   d. jingling keys

8. What is the difference between the two sound waves shown on the oscilloscopes in Figure 27?

   The sound from the French horn is more regular, while the sound from the sharp clap is a mix of random frequencies.

9. When does the amount of noise around you become noise pollution?
   a. when it becomes bothersome
   b. when you can hear it
   c. when it causes health problems
   d. when it is loud

10. According to the Environment Connection on page 555, which of the following groups were NOT affected by the Los Angeles International Airport?
    a. women
    b. men
    c. butterflies
    d. animals

Review (p. 555)
Now that you've finished Section 4, review what you learned by answering the Review questions in your ScienceLog.

Name _____ Date _____ Class _____

CHAPTER 22 DIRECTED READING WORKSHEET

The Nature of Light

As you read Chapter 22, which begins on page 562 of your textbook, answer the following questions.

Strange but True! (p. 562)

1. What causes jaundice?

   The pigment bilirubin builds up in the bloodstream and is deposited in the

   skin.

2. If you were a parent of an infant with jaundice, what method of treatment would you choose? Why?

   Accept all reasonable answers. Sample answer: I would choose the bili

   blanket, because it allows me to treat the condition while still comforting

   the baby.

What Do You Think? (p. 563)

Answer these questions in your ScienceLog now. Then later, you'll have a chance to revise your answers based on what you've learned.

Investigate! (p. 563)

3. Do you think you will be able to tell the difference between fluorescent and incandescent lights using your spectroscope? Why or why not?

   Sample answer: Yes; I will be able to tell the difference because different

   light sources produce different kinds of light.

Section 1: What Is Light? (p. 564)

4. Name at least two things that produce light.

   Sample answer: The sun and electric bulbs are two things that produce light.

Light Is an Electromagnetic Wave (p. 564)

5. Sound, unlike light, requires a _____ medium _____ to travel through.

6. A(n) _____ electromagnetic _____ wave is composed of changing electric and magnetic fields.

Name _____ Date _____ Class _____

Chapter 22, continued

Mark each of the following statements *True* or *False*.

7. __True__ A field can exert a push or a pull on an object.
8. __False__ Fields are made of matter.
9. __False__ Light requires a medium through which to travel.
10. __False__ Light waves are considered longitudinal waves because the electric and magnetic fields are perpendicular to each other.

How Light Is Produced (p. 565)

11. An EM wave is produced by the vibration of a(n) _____ electric _____ field and a(n) _____ magnetic _____ field.

12. _____ Radiation _____ is the emission of energy in the form of EM waves.

Take a moment to look at Figure 2. Then choose the term in Column B that best matches the description in Column A, and write your answer in the space provided. Terms may be used more than once.

Column A	Column B
__a__ 13. move about the nucleus at different distances	a. electrons
__b__ 14. tiny "packets" of energy	b. photons
__a__ 15. negatively charged particles	
__b__ 16. particles that make up an EM wave	

The Speed of Light (p. 566)

17. Light travels _____ faster _____ through a vacuum than through a medium such as air, water, or glass. (faster or slower)

18. Why do you see lightning before you hear the thunder that accompanies it?

    Light travels much faster than sound.

Review (p. 566)

Now that you've finished Section 1, review what you learned by answering the Review questions in your ScienceLog.

ANSWER KEY

Section 2: The Electromagnetic Spectrum (p. 567)

1. What does the bee in Figure 4 see that you can't see?

   Sample answer: Because bees can see ultraviolet light, the bee in Figure 4 can see ultraviolet markings in flowers.

2. Name at least four kinds of EM waves.

   Sample answers: visible light, ultraviolet light, X rays, radio waves, and microwaves

Characteristics of EM Waves (p. 567)

3. You can classify an EM wave if you know its __wavelength__ or __frequency__.

4. The __electromagnetic spectrum__ includes the entire range of EM waves.

5. Take a moment to look at the diagram of the electromagnetic spectrum on pages 568 and 569. We can see most of the EM waves in our world. True or (False) (Circle one.)

Radio Waves (p. 568)

6. When you listen to AM radio, the music you hear is encoded in the radio waves by the variation of the __amplitude__ of the waves, while FM radio waves vary in the __frequency__ of the waves.

Mark each of the following statements AM or FM.

7. __AM__ Waves can travel a longer distance.
8. __FM__ Waves have shorter wavelengths.
9. __AM__ Waves can reflect off the ionosphere.
10. __FM__ Waves can encode more information.
11. __FM__ Music sounds better.
12. __AM__ Waves are used to encode the pictures you see on television.
13. __FM__ Waves are used to encode the sounds you hear on television.

---

14. If you watch cable television, how does the television signal get from a distant broadcast studio to your home? List the steps.

    A wave is transmitted to a satellite.

    A satellite amplifies a signal and relays it back to the ground.

    Waves are detected by ground antennae.

    Waves travel through cables into your home.

Microwaves (p. 570)

Using the information given in Figure 7, fill in the blanks in items 15-18 to complete the description of how a microwave oven works.

15. The microwaves are created by a device called a __magnetron__, which accelerates charged particles.

16. The waves are reflected into the __cooking chamber__ by a metal __fan__.

17. These waves can travel a distance of several __centimeters__ into the food.

18. They cause molecules of water to __vibrate__.

19. Look at the Brain Food on page 570. Name four household items that use radio waves.

    Answers may include garage door openers, baby monitors, radio controlled toys, wildlife tracking collars, televisions, and radios.

20. What does *radar* stand for?

    Radar stands for radio detection and ranging.

21. How are microwaves used? (Circle all that apply.)
    a. detecting heat
    (b.) monitoring airplane movement
    (c.) calculating the speed of a car
    (d.) navigating ships at night

Chapter 22, continued

Infrared Waves (p. 571)

Mark each of the following statements *True* or *False*.

22. __False__ Microwaves have shorter wavelengths than infrared waves do.

23. __True__ Infrared radiation can make you feel warm.

24. __True__ The warmer something is, the more infrared radiation it gives off.

25. __True__ With the right equipment, you can "see" in the dark using infrared radiation.

Review (p. 571)

Now that you've finished the first part of Section 2, review what you learned by answering the Review questions in your ScienceLog.

Visible Light (p. 572)

26. The shortest wavelengths of visible light, which carry the __most__ energy, are seen as the color __violet__.

27. The longest wavelengths of visible light are seen as the color __red__.

28. Look at the Biology Connection on page 572. What kind of EM wave fuels photosynthesis?
    a. microwaves
    b. infrared light
    c. visible light
    d. ultraviolet light

29. Who is Roy G. Biv, and why should you remember his name?

    Sample answer: Roy G. Biv is not a real person, but remembering his name can help me remember the order of the colors in the visible spectrum: red, orange, yellow, green, blue, indigo, violet.

Ultraviolet Light (p. 573)

30. Name two benefits of ultraviolet light.

    Sample answer: Ultraviolet light can be used to sterilize things by killing bacteria, and my skin needs it in order to produce vitamin D.

31. Name two hazards of ultraviolet light.

    Sample answers: It can cause sunburn, skin cancer, damage to the eyes, wrinkles, and premature aging.

X Rays and Gamma Rays (p. 574)

32. How are X rays useful in the medical field?

    X rays allow a doctor to see if bones are broken.

33. Why do you often need to wear a lead apron when getting an X ray?

    The apron protects parts of the body that do not need to be X rayed.

34. Which of the following is NOT true of gamma rays?
    a. They are used to treat cancer.
    b. They carry the least amount of energy of the EM waves.
    c. They penetrate materials easily.
    d. You are exposed to them every day.

Review (p. 574)

Now that you've finished Section 2, review what you learned by answering the Review questions in your ScienceLog.

Section 3: Interactions of Light Waves (p. 575)

1. How is the special layer of cells in the back of a cat's eyes useful?

   This layer reflects light, giving the eyes another chance to detect light.

Reflection (p. 575)

2. When you look in a mirror you see light that has been reflected twice. (**True** or False? (Circle one.)

3. What is the law of reflection?

   The law of reflection states that the angle of incidence is equal to the angle of reflection.

Chapter 22, continued

4. In Figure 14, the __normal__ is the line perpendicular to the mirror's surface.
5. What kind of surface allows you to see your reflection?
   A smooth reflecting surface allows you to see your reflection.

Absorption and Scattering (p. 576)

Decide if each of the following refer to scattering or absorption. Write S for scattering and A for absorption.

6. __A__ EM waves transfer energy to particles in matter.
7. __S__ Particles of matter that have absorbed energy release light energy.
8. __S__ On a dark night, you can see objects outside of a flashlight beam.
9. __S__ The sky appears blue.
10. __A__ Air particles absorb the energy from light, making it less bright.

Refraction (p. 577)

11. Refraction is caused by a variation in the __speed__ of light as it passes from one medium to another.
12. The speed of light traveling through glass is slower than the speed of light traveling through air. (True) or False? (Circle one.)
13. Optical illusions occur because your brain interprets light as traveling in __straight__ lines.
14. Color separation occurs when light is refracted because light with a long wavelength is bent more than light with a short wavelength. True or (False)? (Circle one.)

Diffraction (p. 579)

15. The __bending__ of waves around barriers and through openings is called diffraction.
16. In order for the greatest amount of diffraction to occur, an opening must be the same size or __smaller__ than the wavelength of the light passing through it.

Chapter 22, continued

17. A greater amount of diffraction occurs when light passes through a narrower opening. (True) or False? (Circle one.)

Interference (p. 579)

18. The wave that results from __constructive__ interference has a greater amplitude than the individual waves that combined to form it.
19. The result of __destructive__ interference is dimmer light.
20. Why do you not see constructive and destructive interference of white light?
    White light is composed of waves of many different wavelengths. The
    waves rarely line up to combine in total destructive interference.

Use the information in Section 3 to answer the following questions. Choose the word in Column B that best matches the definition or example in Column A, and write your answer in the space provided.

	Column A	Column B
__c__	21. transfer of energy from light waves to particles of matter	a. interference
__d__	22. bending of light passing through an opening	b. scattering
__e__	23. bending of light passing into a different material	c. absorption
__f__	24. a wave bouncing off an object	d. diffraction
__a__	25. waves overlapping and combining	e. refraction
__b__	26. the release of light energy by particles of matter that have absorbed energy	f. reflection

Review (p. 580)

Now that you've finished Section 3, review what you learned by answering the Review questions in your ScienceLog.

Section 4: Light and Color (p. 581)

1. What is white light made of?
   White light is made of all the colors of light.

Name _____ Date _____ Class _____

Chapter 22, continued

Light and Matter (p. 581)

2. What are the three ways light can interact with matter?
   reflection, absorption, or transmission

3. The three interactions you listed in question 2 can all occur at the same time. (**True** or False? (Circle one.)

In items 4–7, complete each sentence using the word *transparent*, *translucent*, or *opaque*.

4. Most windows are ___transparent___.

5. A frosted glass window is ___translucent___.

6. A wooden door is ___opaque___.

7. An object that is ___translucent___ transmits light but scatters the light as it passes through.

Colors of Objects (p. 582)

8. An object appears a certain color as a result of
   a. the wavelengths of light that reach your eyes.
   b. the wavelengths of light that reach the object.
   c. the size of the object.
   d. the size of the light source.

9. The color of an opaque object is the color of the light that is ___reflected___ by the object. The object ___absorbs___ any remaining colors.

10. What colors of light are reflected by an opaque white object?
    All colors are reflected.

11. What colors of light are reflected by an opaque black object?
    No colors are reflected. The colors are all absorbed.

12. The color of a transparent object is the color of the light that is ___transmitted___ through the object. The object ___absorbs___ any remaining colors.

Mixing Colors of Light (p. 584)

13. The three primary colors of light are ___red___, ___blue___, and ___green___.

Name _____ Date _____ Class _____

Chapter 22, continued

14. Why are the colors you wrote in item 13 called the primary colors of light?
    Sample answer: They can be combined in different ratios to produce all colors of visible light.

15. What happens when you add two primary colors of light together?
    A secondary color of light is produced.

16. The three secondary colors of light are ___cyan___, ___magenta___, and ___yellow___.

Mixing Colors of Pigment (p. 584)

17. How does a pigment give a substance color?
    The pigment absorbs some colors of light and reflects others.

18. Why is the mixing of colors of paint called color subtraction?
    Sample answer: When you mix paint, the mixture absorbs all the light absorbed by the original pigments so more colors of light are subtracted from the reflected light.

19. What are the three primary pigments?
    cyan, magenta, and yellow

20. Take a moment to read the Geology Connection on page 585. What does azurite change into over time?
    Azurite changes into malachite over time.

Review (p. 585)

Now that you've finished Section 4, review what you learned by answering the Review questions in your ScienceLog.

CHAPTER 23 DIRECTED READING WORKSHEET

Light and Our World

As you read Chapter 23, which begins on page 592 of your textbook, answer the following questions.

Imagine . . . (p. 592)

1. Using the retroreflector scientists have learned that the moon is moving ___away from___ the Earth.

What Do You Think? (p. 593)

Answer these questions in your ScienceLog now. Then later, you'll have a chance to revise your answers based on what you've learned.

Investigate! (p. 593)

2. What is the purpose of this activity?

   The purpose of this activity is to investigate the properties of plane mirrors.

Section 1: Light Sources (p. 594)

3. Visible light
   a. covers the entire electromagnetic spectrum.
   b. covers most of the electromagnetic spectrum.
   c. covers a small part of the electromagnetic spectrum.
   d. is not a part of the electromagnetic spectrum.

Light Source or Reflection? (p. 594)

4. Look at the Astronomy Connection on page 594. The sun and the moon are two celestial objects visible from Earth. Which one is luminous, and which one is illuminated?

   The sun is luminous; the moon is illuminated.

DIRECTED READING WORKSHEETS 173

Chapter 23, continued

For each of the following, write *Yes* if it is a light source or *No* if it is not a light source in the space provided.

5. ___Yes___ light bulb
6. ___No___ wooden bookshelf
7. ___Yes___ burning candle
8. ___Yes___ glow-in-the-dark rubber ball
9. ___No___ coin
10. ___Yes___ firefly
11. ___No___ mirror

Producing Light (p. 595)

12. What does an incandescent light emit? (Circle all that apply.)
    a. electricity
    b. thermal energy
    c. light
    d. halogens

13. The inside of a fluorescent light tube is coated with phosphor. What would happen if the phosphor were removed?

    The phosphor produces visible light when struck by ultraviolet light. Without the phosphor, no visible light would be produced. The fluorescent light would give off ultraviolet light only.

In a neon light, the type of gas in the tube determines the color of the light produced. Match the gas in Column A with the color of light in Column B, and write your answer in the space provided.

	Column A	Column B
b	14. neon	a. blue
c	15. sodium	b. red
a	16. argon and mercury	c. yellow

174 HOLT SCIENCE AND TECHNOLOGY

Name _____ Date _____ Class _____

Chapter 23, continued

17. What types of vapor do street lights usually contain?

Street lights usually contain mercury vapor or sodium vapor.

Review (p. 597)

Now that you've finished Section 1, review what you learned by answering the Review questions in your ScienceLog.

Section 2: Mirrors and Lenses (p. 598)

1. A rearview mirror reverses an image left to right. How would the word *LAMB* appear if you viewed it in a rearview mirror? (Hint: look at the figure of the ambulance on the right margin.)

ƎMAJ

Rays Show the Path of Light Waves (p. 598)

2. The path a light wave takes is called a _____ ray because it travels along a _____ straight _____ path. (ray or arc, straight or curved)

Mirrors Reflect Light (p. 599)

3. Imagine that Sarah, a new college student, goes to the bookstore and buys herself a sweatshirt with the name of her college on it. Back in her dorm room, she puts on the shirt and looks in the mirror (a plane mirror). Describe the image she sees.

She would see an upright, virtual image of herself. The name of the college

would be reversed left to right.

Concave mirrors produce different types of images depending on where the reflected object is placed. Match the object placement in Column A with the image created in Column B, and write your answer in the space provided.

Column A	Column B
_b_ **4.** in front of the focal point	**a.** real image
_c_ **5.** at the focal point	**b.** virtual image
_a_ **6.** beyond the focal point	**c.** no image

Name _____ Date _____ Class _____

Chapter 23, continued

7. Imagine you are at a carnival funhouse looking at a mirror. No matter how far you stand away from the mirror, your image is upright and small. What type of mirror are you looking at?

Answers may vary. Sample answer: I am looking at a convex mirror. Convex

mirrors always form virtual, upright, small images.

Review (p. 602)

Now that you've finished the first part of Section 2, review what you learned by answering the Review questions in your ScienceLog.

Lenses Refract Light (p. 603)

8. How are lenses classified?
- **a.** by their color
- **b.** by the type of light they refract
- **c.** by their size
- **(d.)** by their shape

9. If you shined two laser pointers in parallel at a convex lens, the lens would _____ refract _____ the light and the beams would cross at the _____ focal point _____. (reflect or refract, optical axis or focal point)

10. Why can't you form a real image with a concave lens?

Concave lenses refract parallel beams away from each other, so the beams

can never cross and form an image.

Review (p. 604)

Now that you've finished Section 2, review what you learned by answering the Review questions in your ScienceLog.

Chapter 23, continued

Section 3: Light and Sight (p. 605)

How You Detect Light (p. 605)

1. What controls the size of the pupil?

   The iris controls the size of the pupil.

2. In bright light the muscles of the iris contract, making the pupil _smaller_ (larger or smaller).

3. In low light the muscles of the iris relax, making the pupil _larger_ (larger or smaller).

4. What do the muscles surrounding the lens do?

   The muscles change the thickness of the lens to focus on nearby or far away objects.

Common Vision Problems (p. 606)

Imagine that two of your friends, Emily and Harold, left their glasses at your house. Emily is nearsighted. Harold is farsighted. One of the pairs of glasses has convex lenses, the other has concave lenses. Match the description in Column A with the friend in Column B, and write your answer in the space provided. Answers can be used more than once.

Column A	Column B
_b_ 5. wears convex lenses	a. Emily
_a_ 6. wears concave lenses	b. Harold
_a_ 7. eyes are too long	
_b_ 8. eyes are too short	

9. What specific problem would a driver who is red-green colorblind face?
   a. reading stop signs
   b. seeing double
   c. distinguishing between stop and go lights
   d. checking his or her blind spot

10. Explain why there are not corrective lenses for colorblindness as there are for nearsightedness and farsightedness.

    Lenses move the focal point of light entering the eye, but they cannot affect how cones in the retina react to that light.

Review (p. 607)

Now that you've finished Section 3, review what you learned by answering the Review questions in your ScienceLog.

Section 4: Light Technology (p. 608)

Optical Instruments (p. 608)

The camera is an optical instrument that is similar in some ways to your eye. Match the camera parts in Column A with the corresponding eye structure in Column B, and write your answer in the space provided.

Column A	Column B
_b_ 1. aperture	a. retina
_d_ 2. shutter	b. pupil
_c_ 3. lens	c. cornea and lens
_a_ 4. film	d. no corresponding structure

5. An astronomer is looking at light from a distant star through a refracting telescope. Below is a list of items the light ray would encounter between the star and the astronomer's retina. In which order would the light ray strike or pass through them? Write the items in order in the spaces provided, and indicate if the item would refract or reflect the light ray.

   telescope objective lens

   cornea and lens of astronomer's eye

   telescope eyepiece lens

   telescope objective lens; refracted

   telescope eyepiece lens; refracted

   cornea and lens of astronomer's eye; refracted

Name _____ Date _____ Class _____

Chapter 23, continued

6. Another astronomer is looking at light from the same distant star through a reflecting telescope. Below is a list of items the light ray would encounter between the star and the astronomer's retina. Write the items in order in the spaces provided, and indicate if the item would refract or reflect the light ray.

cornea and lens of the astronomer's eye
telescope plane mirror
telescope eyepiece lens
telescope concave mirror

telescope concave mirror; reflected

telescope plane mirror; reflected

telescope eyepiece lens; refracted

cornea and lens of astronomer's eye; refracted

7. Light microscopes are similar in construction to refracting telescopes.

(refracting telescopes or reflecting telescopes)

Lasers and Laser Light (p. 609)

8. Unlike white light, laser light is _____ coherent _____, meaning that all the light waves move together away from the source.

9. Using Figure 23, put the following items in the correct order to show how a laser is created.

   2 Excited neon atoms release photons, which strike other atoms.

   4 Some photons leak out of a partially coated mirror to create a beam.

   3 Plane mirrors on the ends of the laser reflect photons back and forth.

   1 An electric current excites neon atoms in a gas-filled tube.

10. What causes the laser light in Figure 23 to become brighter?

Because photons travel back and forth many times, many stimulated emissions occur, making the laser light brighter.

Name _____ Date _____ Class _____

Chapter 23, continued

11. What would happen to the hologram being made in Figure 24 if the beam splitter were removed?
    a. This would have no effect.
    b. The laser would quit shining due to the interference pattern.
    c. There would be no interference pattern formed on the film.
    d. The object would melt.

12. What common household device contains a laser?

CD players (and laser pointers) have a laser in them.

Fiber Optics (p. 612)

13. Why have fiber-optic cables largely replaced copper telephone cables?

Fiber-optic cables transmit data faster and more clearly than copper cables.

14. Light cannot escape an optical fiber because of total _____ internal reflection _____.

Polarized Light (p. 612)

15. Polarizing sunglasses decrease _____ glare _____ from horizontal surfaces by reflecting light that vibrates horizontally.

16. The two photos in Figure 28 were taken by the same camera from the same angle. Why are they different?

There is less reflected light in the photo at right because a polarizing filter was placed over the lens of the camera.

Review (p. 613)

Now that you've finished Section 4, review what you learned by answering the Review questions in your ScienceLog.